ECOLOGY

T.J. King

SECOND EDITION

T. J. King M.A. D.Phil. (Oxford), F.I.Biol.

Director of Studies, Abingdon School

Thomas Nelson and Sons Ltd
Nelson House Mayfield Road
Walton-on-Thames Surrey
KT12 5PL UK

Nelson Blackie
Wester Cleddens Road
Bishopbriggs
Glasgow
G64 2NZ UK

Thomas Nelson Australia
102 Dodds Street
South Melbourne
Victoria 3205 Australia

Nelson Canada
1120 Birchmount Road
Scarborough Ontario
M1K 5G4 Canada

© T J King 1989

First published by Thomas Nelson and Sons Ltd 1989

I(T)P Thomas Nelson is an International
        Thomson Publishing Company

I(T)P is used under licence

ISBN 0-17-438404-1
NPN 9 8

Printed in Croatia

Cover photograph shows a spider (*Argiope bruenichi*). Courtesy of Tony
Stone Worldwide.

Preface to the First Edition

Over the last ten years ecology, human population growth, pollution and conservation have crept more and more into sixth-form teaching. This book provides a brief, modern account of pure and applied ecology. I hope that it will be read by sixth formers as part of their A-level studies.

The first chapter is an introduction to the subject. The next five chapters are concerned with the ecology of individual species. The concept of the niche, the factors which determine the distribution patterns of species, and population dynamics are discussed in turn. Chapters seven and eight deal with the interactions of species in communities. A brief account of succession is followed by a detailed description of the ecosystem concept. The last two chapters summarise pollution and conservation. The book ends with a course of class practicals and some advice on the planning and execution of individual projects.

Any budding ecologist must have first-hand acquaintance with organisms in the field. It is imperative that those who read this book should be participating in a course of field observations and laboratory exercises at the same time.

I am conscious that ecology is difficult to generalise about and dangerous to simplify. Those already well versed in the subject will notice that I have omitted some important concepts, examples and technical terms. I have done this in the interests of clarity, comprehensibility and readability. My aim has been to erect a framework of principles.

T. J. King

Preface to the Second Edition

The first edition is widely used as an A-level text. I have retained the concise summary of principles and the emphasis on experiments, which are even more relevant now than they were ten years ago. In this edition, however, I have extended and re-organised the sections on applied ecology, expanded the practical suggestions at the end of the book, provided advice on projects, and brought the text and references up to date. The intention is to provide a text which, although unashamedly an account of modern theoretical ecology, can be read with profit by post-GCSE students and their teachers and can be used as a source of ideas in practical and project work.

T. J. King
September 1989

Contents

Acknowledgements

I am most grateful to my wife Heather, who kept the children at bay whilst I wrote both the first and second editions of the book.

Michael Roberts' detailed and perceptive comments on the first edition considerably improved the text. This was also knocked into shape by Dr Stanley R.J. Woodell (Botany School, Oxford University), Tony Hollander (Magdalen College School, Oxford) and Wayne Masterson (ex-Magdalen College, Oxford). Dr John Phillipson (Zoology Department, Oxford University) improved the chapter on ecosystems. Many ideas were unwittingly contributed by my former school and university pupils, particularly Simon May, Humphrey Birley, Pat Frew, Simon Hayward and John Goodchild.

I am also most grateful to Elizabeth Johnston and Donna Evans for their help and encouragement during the production of the first edition, and Chris Coyer and Nicky Benton of Nelson's for engineering the second edition.

Acknowledgement is due to the following for permission to use copyright material as a basis for some of the illustrations in this book

fig. 4, page 5   H. Godwin, (1956) *The History of the British Flora.* (Cambridge University Press.)   fig. 11, page 16   G. L. Hodgson and G. E. Blackman, *J. Exp. Bot., (1956)*, 7, 146.   fig. 16, page 25   J. H. Connell, *Ecology, (1961)*, 42   fig. 24, page 38   Gerald Thompson, Oxford Scientific Films   fig. 30, page 49   G. E. Hutchinson (1978) *An Introduction to Population Ecology.* (Yale University Press.)   fig. 32, page 51   C. M. Perrins (1979) *British Tits.* (Collins, Sons & Co.)   Shelford, *Auk, (1954)*, 62, 592 appearing in D. Lack (1954) *The Natural Regulation of Animal Numbers.* (Oxford University Press.) fig. 36, page 52   E. P. Odum (1971) *Fundamentals of Ecology*, 3rd edn. (W. B. Saunders Co (Philadelphia).)   fig. 37, page 54   G. C. Varley, G. R. Gradwell and M. P. Hassell (1973) *Insect Population Ecology.* (Blackwell Scientific Publications.)   fig. 40, page 62   G. T. M. Hirons and K. T. Marsland, Animal Ecology Research Group. Department of Zoology, Oxford.   fig. 46, page 71   D. H. Cushing (1968) *Fisheries Biology* (University of Wisconsin Press.)   fig. 48, page 76   E. S. Deevey, Jr. *Scientific American, (Sept 1960)*

(© Scientific American, Inc. All rights reserved.) fig. 51, page 79 James
Echols. Population vs the Environment. *American Scientist March 1976*, **64** fig.
54, page 87 J. Sarukhan, *J Ecol, (1974)*, **62**, 151 (British Ecological Society)
fig. 55, page 87 J. Sarukhan, J. L. Harper *J Ecol, (1973)*, **61** 675 (British
Ecological Society) fig. 61, page 101 H. Walter (1973) *Vegetation of the Earth
in relation to Climate and the Ecophysical Conditions.* (Springer-Verlag, New York,
Inc.) fig. 65, page 106 H. W. Harvey (1955) *The Chemistry and Fertility of
Sea water* (Cambridge University Press.) fig. 70, page 113 C. J. Krebs
(1978) *Ecology — The Experimental Analysis of Distribution and Abundance.* (Harper
and Row, New York.) Appearing in Open University Course S323 (1974)
fig. 7, page 123 Paul R. Ehrlich and Anne H. Ehrlich (1972) *Population
Resources, Environment: Issues in Human Ecology.* 2nd Edition (W. H. Freeman
and Co, San Francisco.)

Acknowledgement is due to the following for permission to reproduce
photographs.

fig. 1, page 3; fig. 3, page 4; fig. 12, page 18; fig. 22, page 35; S. R. J. Woodell,
Botany School, Oxford. fig. 2, page 3 E. V. Tanner, *Journal of Ecology
(1977).* fig. 8, page 10 G. D. Kinns. fig. 9, page 13 Richard Maughn,
Ardea. fig. 15, page 22 Heather Angel. fig. 23, page 37 John Clegg,
Ardea. fig. 25, page 39 Heather Angel. fig. 29, page 45 Tim King.
fig. 35, page 52 Edgar James, Ardea. fig. 40, page 62 G. T. M. Hirons
and K. T. Marsland, Animal Ecology Research Group. Department of Zoology,
Oxford. fig. 41, page 63 C. M. Perrins (1979) *British Tits.* (Collins, Sons
& Co.) fig. 43, page 68 Professor P. de Bach, University of California
fig. 47, page 74 Guildhall Library, City of London. fig. 60, page 98
Rothamsted Experimental Station, Harpenden, Herts. fig. 67, page 109
© Douglas P. Wilson. fig. 84, page 133 Photograph by Gene E.
Likens. fig. 88, page 143 National Parks Service, U.S.A. fig. 87, page
141 Edgar James, Ardea. fig. 89, page 145 Imperial Chemical Industries.

Units and Symbols

> greater than
< less than

Prefixes

T	=	tera	= 10^{12}
G	=	giga	= 10^9
M	=	mega	= 10^6
K	=	kilo	= 10^3
d	=	deci	= 10^{-1}
c	=	centi	= 10^{-2}
m	=	milli	= 10^{-3}
μ	=	micro	= 10^{-6}
n	=	nano	= 10^{-9}

Suffixes

°C	=	degrees Celsius (0°C = 273 K)
J	=	Joule (= 0.239 cal)
m	=	metre
s	=	second
h	=	hour
g	=	gram
mole	=	6.023×10^{23} particles (ions, atoms or molecules)
l	=	litre
ha	=	$10^4\,m^2$
ppm	=	parts per million
spp.	=	species (plural)
atm	=	atmosphere
N	=	Newton
bar	=	$10^5\,N\,m^{-2}$

1 Introduction: The scope of ecology

Ecology, a word derived from the Greek for 'house', is the scientific investigation of animals, plants and micro-organisms in their natural surroundings. It encompasses the study of the distribution patterns of organisms, their numbers, their interactions with one another, and their relationships to their environment. Ecology is therefore a large and important segment of modern biology.

Yet ecology is not only a fascinating and enriching enquiry. It is essential to the human race. Our supplies of food, oxygen and natural products are derived from areas of modified wilderness, and depend on the activities of other organisms. If these supplies are to be maintained in the long term, in the face of the rapidly increasing human population and its industrial wastes, the more we know about other organisms, the better.

The scope of ecology is therefore vast. One ecologist could be trying to understand the cycling of the element nitrogen over the surface of the earth. Another might be studying the small animals found in the leaf bases of a certain species of moss. The ecological literature consists of hundreds of thousands of publications on individual organisms and communities, most of them without a common purpose. Nevertheless, some general principles are emerging and it is on these that I concentrate.

This chapter introduces the basic principles and terms. It begins with a discussion of the levels on which organisms can be studied and how an ecologist approaches a problem. Then an oak wood is analysed from an ecologist's point of view.

THE LEVELS ON WHICH ORGANISMS CAN BE STUDIED

The earth is a lump of rock shooting through space, with a thin rind of soil, water and air. Most organisms occur near the boundaries between soil and air on the one hand, or water and air on the other. These organisms can be studied at six different levels, the biosphere, biome, ecosystem, community, population and individual, each level being a subset of the previous one. Let us consider them in turn.

(a) The volume of the earth's surface in which organisms can be found is known as the **biosphere**. It extends from the depths of the deepest oceans (11 km below sea level) to at least the highest plant communities (6.2 km).

(b) The biosphere is composed of several types of **biome**, which are major vegetation types such as the arctic tundra, the tropical rain forest and the sea. Biomes differ from one another in the structure of the vegetation, the sorts of plant and animal species they contain, their geographical ranges and the climatic regimes under which they exist.

Tundra, for example, occurs at high latitudes and on the tops of mountains, in regions of cold climates and short growing seasons. It is treeless and is dominated by shrubs of the heather family, mosses and lichens (figure 1). The plants grow slowly, food chains are short and there are few animal species. In contrast, tropical rain forests occur in moist environments near the tropics. There is intense sunlight all year round, although the atmosphere is usually cloudy or hazy. The plants grow rapidly, the tall evergreen trees are festooned with smaller plants like orchids growing on their trunks and branches, and beneath the trees there is a lush undergrowth of ferns and shrubs (figure 2). Food chains are long, there are myriads of species, and the forests are filled with the noisy sounds of amphibians, birds and mammals.

(c) Whole biomes, however, are difficult to study because they extend over such large geographical areas. **Ecosystems** are usually parts of biomes and have attracted a great deal of attention. Each ecosystem, like a pond or a forest, has a characteristic set of plants, animals and micro-organisms. These organisms influence, and are influenced by, the climate and the soil and form, usually, a self-sufficient unit in balance with its environment. An oak wood is discussed from the ecosystem point of view later in this chapter (p. 9), and ecosystems are dealt with in detail in chapter 6.

(d) At a lower level still, a **community** is a group of species which occurs in the same place at the same time. The word is often used to refer to organisms of a particular kind, such as the plant community on a lawn or the insect community inside a bracket fungus on the trunk of a tree.

(e) Ecosystems and communities contain **populations** of species. A population is made up of all the members of the same species which occur in the same place at the same time, like the sticklebacks in a stream or the ash trees in a wood.

(f) Finally, each species population is made up of many **individuals**. The genetical and physiological adaptations of an individual organism to its environment are an important aspect of ecology.

Ecologists study organisms at each of these six levels. Most ecologists, however, concentrate either on individual species, or on ecosystems and communities. The study of individual species is called autecology and is discussed in the first four chapters. Research into groups of species, in com-

FIGURE I

FIGURE 2

Arctic tundra at Godhavn, Greenland (Latitude 70°N), with the Arctic Station of the University of Copenhagen in the background. The hummocks in the foreground were formed by frost-heaving and are covered by plants from the heather family. (S. R. J. Woodell.)

Dwarfed rain forest in the western part of the Blue Mountains of Jamaica (Latitude 18°N), at about 1550 in altitude. Tree ferns are particularly obvious. (Courtesy of E. V. J. Tanner).

munities and ecosystems, is known as synecology and is dealt with in chapters 5 and 6.

WHAT DO ECOLOGISTS DO?

Ecologists are not merely naturalists, but research scientists with a solid grounding in meteorology, geology, soil science, chemistry and physics. A naturalist might merely look for particular species and record the events he witnesses. An ecologist tries to see patterns, seeks an explanation for them, tests his ideas by experiment and records numerical data.

Suppose, for example, that a farmer calls in a veterinary surgeon because his sheep are not putting on weight. The vet, having read of similar cases, might suspect pollution from the nearby aluminium smelting factory. A close examination of the area by an ecologist might reveal that mosses and liverworts are absent, no lichens grow on the fence posts, and the leaves on birch (*Betula* spp.) and willow (*Salix* spp.) are dying in August. The idea, or hypothesis, that these unusual events are due to pollution needs to be tested by experiment.

The ecologist would analyse air, soil, vegetation and the bones of dead sheep from a site near the factory. Another site, similar in all respects except that an aluminium smelter was absent, would be selected for comparison as the 'control' area. The analyses might show a much higher level of fluoride near the factory than in the 'control'. It is well known that the cells of mosses and lichens absorb fluorides readily and that several important enzyme-controlled reactions in the cytoplasm are disrupted by fluorides.

Mosses and lichens could be transplanted into the area of the factory and the control area. Their progress could be followed, and after some time they could be analysed for fluoride. Their symptoms might be compared with those of mosses and lichens exposed in the laboratory to a range of fluoride concentrations. In this way it could be established beyond reasonable doubt that the release of fluoride compounds from the factory was affecting the organisms nearby.

The same sort of experimental approach can be applied to any ecological problem. As an ecologist wanders around the landscape, many questions occur to him. Possible answers spring to mind, hypotheses which could be tested by experiment. Let us illustrate this by considering an oak wood.

FIGURE 3

An ecologist at work. Hal Mooney, of Stanford University, measures the net photosynthesis of bristlecone pine (*Pinus aristata*) in the field with an infra-red gas analyser. Bristlecone pines are the longest-lived organisms on earth; the oldest felled tree was 4600 years old. (S. R. J. Woodell.)

FIGURE 4

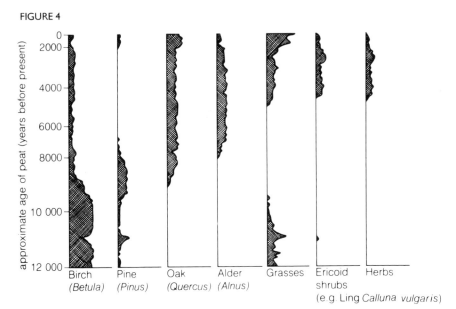

Simplified pollen diagram for Hockham Mere, in the north of the Breckland of East Anglia. Pollen grains were extracted from peat of different known ages. After the ice age (bottom of graph) birch and pine predominated in the area. Oak began to dominate about 8000 years ago. Herb, heather (*ericoid shrubs*) and grass pollen appeared 4500 years ago as these plants invaded land cleared of forests by humans. (Modified from H. Godwin (1956). *The History of the British Flora*. Cambridge University Press.)

AN ECOLOGIST'S VIEW OF AN OAK WOOD

Imagine a typical oak (*Quercus robur*) wood in southern England. How old is it? Oak forest is a **climax** community, that is, it is self-perpetuating. When one tree dies another replaces it. The wood could persist apparently unchanged for thousands of years. The pollen grains of oak first occur abundantly in Britain in peat about 8000 years old. Oak has been a major feature of the pollen record ever since (figure 4). Thus the maximum possible age of the wood, in the absence of considerable climatic change or drastic human interference, is 8000 years.

In most woods, however, human influence has been considerable. Woods have provided timber for erecting houses, furniture, ships and fences. They also produce firewood, charcoal, and grazing for many domesticated animals. Information about the history of the wood could be obtained from old maps, or parish and estate records. Are the trees planted, or self-sown? Indications that they had been planted might be the absence of seedlings, a regular pattern, and a similar age and size. How old are the trees? A thin tube cut from the trunk of a tree enables the annual rings to be counted. Incidentally, a comparison of the widths of different years' rings in old wood may provide valuable information about climatic fluctuations in the past.

In one area of the wood the oaks are smaller and younger, mixed with some ash and hawthorn. Could there have been human disturbance? At one time, up to a hundred years ago, this site was probably a pasture or an arable field. Then it was abandoned, and a **succession** took place. A succession is a gradual, predictable change in the plant and animal species at a site, leading eventually to the establishment of a climax community like oak woodland (see chapter 5). In this case, weeds of cultivation probably became established first on the abandoned arable field. Shrub seedlings and tree seedlings began to appear, and the area became scrub and then woodland.

Why do bluebells (*Endymion non-scriptus*) and wood anemones (*Anemone nemorosa*) grow beneath the trees whilst grassland plants like daisies (*Bellis perennis*) and plantains (*Plantago* spp.) do not? Bluebells overwinter as bulbs and wood anemones have rhizomes (underground stems). They both produce leaves very early in Spring. By the time that the trees establish a dense leaf canopy in late May or early June, bluebells and wood anemones have accomplished most of their photosynthesis. In contrast, daisies and plantains photosynthesise too slowly, and respire too rapidly at low light intensities, to survive under a woodland canopy. Furthermore, when their seeds are sown in woodland their seedlings rarely survive fungal attack.

A casual observer might notice little else but oak trees in the wood. He might hear some distant bird song, see the flies buzzing round his head and notice some bees visiting flowers.

Most animals are hidden from view by what Charles Elton has called the 'curtain of natural cover'. An experienced ecologist would see a bewildering variety of organisms. He might examine the undersides of leaves, turn over some stones or go out at night with a torch. In a twelve-year study of a group of twenty-one oak trees at Wytham Woods near Oxford, more than a thousand insect species were collected.

All its life an oak tree will influence, and be influenced by, other organisms. The acorn from which it grew was one of the few which was not eaten by jays, pigeons, squirrels, rabbits or pigs. It germinated in the leaf litter produced by trees and shrubs, in the partial shade and humid conditions created by the trees around it. Its roots may be infected by mycorrhizal fungi, which help it to absorb the nutrient ions released when earthworms, springtails, mites, bacteria and fungi break down leaf litter. The oxygen molecules which it uses in cellular respiration may have been released in photosynthesis by the leaves of the surrounding trees, and some of the carbon dioxide used in photosynthesis may have been produced in the cellular respiration of any of the organisms in the wood. As it grows, the oak tree may be sat on, excreted on and nested on by birds; pricked and deformed by gall-forming wasps and chewed by their larvae; sucked by aphids; defoliated by the larvae of moths; or attacked by fungi. It may be used as a platform by mosses and lichens, and as a climbing frame by ivy. Underneath the oak will grow characteristic plants, and various animal species will use the leaves, twigs, branches, trunks and roots for cover, shelter, and food.

The tree is like a house with many rooms, each offering a different meal and a different environment. From the trunk to the edge of the tree, and from the top to the bottom, the environment changes; the upper rooms have central heating but the cellars are cold and damp. In each room several jobs are available for organisms to perform.

As the external climate changes from month to month, the climate in each 'room' in the oak tree will change. In autumn the tree loses its clothes and will stand naked all winter. The microclimate in its leaves and branches will alter abruptly. The soil organisms receive an annual windfall of dead leaves.

The ecology of individual species in the wood

The animals will only stay in the wood if they find their food on or near the tree and if conditions there are suitable for activity, mating and the development of the young. For these species the oak wood is their **habitat**, the place where they live. Some of the factors which restrict species to particular habitats on a geographical and local scale are discussed in chapter 3.

Most species can be found in a specific part of the tree. The larvae of the winter moth (*Operophtera brumata*), for example, are found feeding on the buds at the tips of oak twigs during April and May. The buds are the **microhabitat** of the larvae at this time of year, in other words, the place where they live, considered on a small scale.

Besides a habitat and a microhabitat, every species has a job or role in the community, known as its **niche**. The niche of a species includes its diet ('feeding niche'), the micro-environment in which it finds its food ('structural niche') and the way in which it gathers is food ('behavioural niche').

The feeding niches available in an oak wood are listed in figure 5. The niche of the winter moth in oak woodland is to graze the leaf buds of oak trees and to provide food for a wide range of parasites, birds, and mammals. In ecology, the niche of an organism is its profession and its microhabitat is its address. Niches are discussed in chapter 2.

Apart from niche, microhabitat, habitat, distribution and life-cycle, ecologists also investigate the population dynamics of individual species. Population dynamics, dealt with in detail in chapter 4, includes the study of the fluctuations in population size of a species from year to year, the causes

FIGURE 5

The feeding niches in an oak wood. (After Imms and E. Neal (1953). *Woodland Ecology.* Heinemann, London.)

(a) Herbivores
Feeders on leaves, buds and shoots
  i defoliators
  ii miners
Suckers of plant juices
  i phloem feeders
  ii nectar feeders
Feeders on fruits and seeds
Formers of galls
Feeders on bark
Borers of wood
Feeders on roots
Feeders on fungi

(b) Carnivores
Predators
Blood suckers
Parasites

(c) Omnivores

(d) Detritivores
Feeders on dead plant matter
Feeders on dead animal matter
  i dung feeders
  ii feeders on dead animals

FIGURE 6

Food web for Wytham Wood, near Oxford, showing the position of the winter moth. (Adapted from G. C. Varley (1970). The concept of energy flow applied to a woodland community. In: A. Watson, (ed.) Animal population in relation to their food resources. *Symposium of the British Ecology Society* **10**, 389–405.) (Blackwell Scientific Publications.)

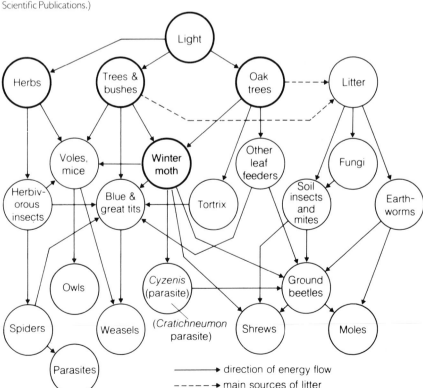

FIGURE 7
Herbivore an animal which eats living plants.
Carnivore an animal which eats living animals.
Omnivore an animal which eats living (and sometimes dead) animals and plants.
Predator an animal which chases, captures, kills and devours other animals (its prey). Predators are carnivores*.
Parasite an organism which forms a semi-permanent association with a larger organism of a different species (its host). Parasites live in or on their hosts during part of the parasite life cycle. They obtain from their hosts carbon compounds which the host would otherwise use.

Saprotroph an organism which attacks dead organisms and excreta, releasing enzymes onto its food and absorbing the soluble breakdown products of digestion.
Decomposers saprotrophic micro-organisms, such as fungi and bacteria.
Detritivore an animal which eats dead organisms or excreta.

(*The word predator has recently been used to refer to herbivorous animals. For instance insects which eat fruits and seeds have been called 'seed predators'. This use of the term has been avoided in this book because it conflicts with the use of the word in everyday speech.)

Definitions of the terms which are used to describe what organisms feed on.

of death, and the factors which tend to keep the population size constant. The winter moth was studied over twelve years on five oak trees at Wytham Woods. The larvae on the buds were eaten by birds, the pupae in the ground were predated by shrews and ground beetles, and the moth was parasitised by two insects and a protozoan. It was the eating of pupae by shrews and ground beetles which seemed to stabilise the population size from year to year. In fact the winter moth is an important component of the food web in the oak wood (figure 6).

Food webs in the oak wood
Oak supports a much greater variety of animal species than any other plant in Britain. There are at least 284 plant-eating insects specifically associated with it, including eighty-two moths. These insects are eaten by many different spiders, harvestmen, insects, birds and mammals. The basic terms which are used to describe what organisms eat are defined in figure 7.

A **food web** is a diagram which shows what each species in a community eats. The simplified food web for a group of oak trees in Wytham Woods (figure 6) illustrates the complexity of feeding interactions in nature. Arrows go from the prey to the predator or from the host to the parasite. In other words, the arrows point in the direction of energy flow. A single sequence of feeding interactions taken from a food web, say oak → winter moth → shrew, is called a **food chain**. It is usually an over-simplification. In

FIGURE 8

The wood mouse (*Apodemus sylvaticus*), a ubiquitous small rodent of woodland throughout the British Isles. It is an omnivore. In autumn and winter it eats mainly seeds. In spring and summer it eats caterpillars, centipedes and other arthropods. Its food also contains seedlings, buds, snails, fungi, moss, bark, galls, dead leaves and earthworms. Wood mice are themselves eaten by many different mammals and birds. (Courtesy of G. D. Kinns.)

this example, winter moths eat blackthorn, hazel, birch, hawthorn and willow besides oak, and shrews eat a wide variety of animal foods. A food web for the whole community of over two thousand species of insects, birds, and mammals in the wood would be difficult to understand. To simplify the information, each organism can be put into one of several **trophic levels** (= feeding levels, since *trophos* is Greek for food). All the organisms at a particular trophic level can be considered together, and their numbers, their mass, or their energy uptake can be totalled.

Just as in human society, there are producers and consumers. The green plants are called producers since they produce the food on which the other organisms depend. The herbivores, like the winter moth, are primary (1st) consumers. They are eaten by secondary consumers, like ground beetles, which are eaten in turn by tertiary (3rd) consumers, and so on. Of course species at the secondary, tertiary and higher consumer levels are all carnivores.

Dead organisms, dead parts of organisms and the excreta of animals shower onto the woodland floor. They form a layer of dead organic matter called detritus. This may be eaten and digested by animals called detritivores, such as earthworms, many species of slug, springtails and dung-flies. These detritivores have their own parasites and predators and so many support an additional food web in the soil. Detritus may also be attacked by micro-organisms like fungi and bacteria, which are known as 'decomposers'. They secrete enzymes onto their food and absorb the soluble breakdown products of digestion. The dead remains of detritivores and decomposers are attacked by living detritivores and decomposers.

All the organisms in the wood need energy and nutrient ions. Thus the rates at which energy and nutrient ions are transferred from one trophic level to the next are of considerable interest.

Energy flow in the oak wood

The solar energy which keeps all the organisms in the wood alive is captured by the millions of photosynthetic chloroplasts in the cells of the oak leaves. This energy is trapped in molecules like glucose. The formation of compounds like glucose increases the dry mass of the oak tree, since glucose is made out of carbon dioxide from the air and water from the soil.

The energy and dry mass which this store of glucose represents is used by the trees in two main ways. Some of it is lost in chemical reactions within the cells, including cellular respiration. The rest is stored in organic carbon compounds and represents potential food for the other inhabitants of the wood. The rate at which the producers accumulate this energy store is known as the **net primary productivity** of the ecosystem (see p.91).

In an oak–ash–sycamore wood at Wytham Woods near Oxford the net primary productivity is about $26 \, MJ \, m^{-2} \, year^{-1}$. Where does this energy go?

In a wood much less energy passes through the grazing food web, i.e. plants→herbivores, than through the detritus food web, i.e. plants→detritivores and decomposers. At Wytham only 3.4% of the net primary production enters the mouths of herbivores. Most of the rest, over 88%, falls on to the soil as leaves, flowers, acorns, twigs and branches. There it begins to decompose under the influence of organisms. The other 8% of the net primary production is stored in dead wood and will eventually reach the detritivores and decomposers when the trees die.

Thus, to sum up, short wave radiation from the sun provides the energy which is trapped in chemical bonds by green plants. Some of this energy is used to do mechanical and chemical work, enabling the organisms to move, feed, reproduce, grow, excrete and so on. Eventually all the chemical bond energy is degraded to heat. This is lost to outer space as long-wave radiation, and cannot be used by other organisms to do useful work. Energy therefore flows through the ecosystem; it goes in at one end and comes out at the other.

Apart from energy flow, the other major aspect of an ecosystem is nutrient cycling.

Nutrient cycling in the oak wood

The nutrients are the chemical elements which are essential to plants and animals. Since the nutrients released from one organism are eventually absorbed and used by another, nutrients cycle round and round in the ecosystem.

Some elements like potassium and calcium, remain as ions both in the soil and inside organisms. Others, such as carbon, hydrogen, oxygen, nitrogen, sulphur and phosphorus, have more complicated cycles. After they have been absorbed by plants, they are incorporated into a variety of carbon compounds and continually appear in new guises. Let us, for simplicity, consider the cycling of potassium ions in the oak wood.

The potassium ions absorbed by the oak trees from the soil will eventually reach the soil again. Some will be rapidly washed out of the leaves. Some potassium ions will be locked up in the leaves and twigs which fall down in autumn. Some will pass up the grazing food web, through primary, secondary and tertiary consumers, and will reach the soil surface when organisms die, urinate or defaecate. The rest of the potassium ions, in the trunk and the roots, will only become available to the detritivores and decomposers when the tree dies.

The cycle is completed when detritivores and decomposers break open cell walls and cell membranes, and the potassium ions are washed into the soil to be absorbed again by the roots of the trees.

The cycle is not completely closed. Some ions will be washed down into the water table and will eventually disappear in rivers and streams. This loss will be balanced by the gain from the weathering of bedrock and soil particles, and from rainfall. Despite this turnover, most potassium ions will reside in the oak wood for tens of years and may pass through thousands of individuals during that time.

Energy flow and nutrient cycling in ecosystems are discussed more fully in chapter 6. Before dealing with the complex interactions of species in ecosystems, however, we shall concentrate for the first part of the book on the ecology of individual species.

2 | *Niches and competition*

The concepts of the niche and competition permeate any discussion of the distribution pattern and numbers of a species. As an example of this, consider the cormorant (*Phalacrocorax carbo*). The structural niche of this bird is broad cliff ledges and the sea. Its feeding niche is that of a bottom feeder, taking mainly flatfish, prawns, shrimps and gobies from shallow estuaries and harbours.

The shag (*P. aristotelis*) has a similar life-style. It nests on cliffs and dives into the water for fish (see figure 9). It is impossible to discuss the distribution limits or population size of the cormorant without considering the possibility that the shag competes with it for nest sites or food. The more their niches overlapped, the more the species would obtain the same resources, and the more the distribution and abundance of one species would be affected by the other.

FIGURE 9

Shags (*Phalacrocorax aristotelis*).

In fact, the shag nests on narrower cliff ledges, feeds further out to sea, and captures fish and eels from the upper layers of the water. The cormorant and shag, like many pairs of closely-related species, differ in ways which tend to reduce competition between them. Nevertheless, it is quite possible that the structural and feeding niches of the cormorant would be far broader in the absence of the shag, and vice-versa. As Charles Darwin described it: '[Some biologists make] the deeply-seated error of considering the physical conditions of a country as the most important for its inhabitants; whereas it cannot, I think, be disputed that the nature of the other inhabitants, with which each has to compete, is generally a far more important element of success.'

THE NICHE

The niche of a species is its role in the community (p.7). A precise and widely-accepted way to define the niche of a species was suggested by G. Evelyn Hutchinson in 1957. Imagine that each factor important in defining the niche of an animal species, such as foraging height, or the sizes of food-stuffs eaten, is represented by a different axis on the same graph. The niche of the species can then be represented as an area on the graph, as shown for the blue–grey gnatcatcher in figure 10. The axes are known as niche dimensions, and the area inside the figure on the graph is called niche space. A third axis, say light intensity, could be added to the graph, in which case the niche would be represented by a three-dimensional figure like a sphere.

FIGURE 10

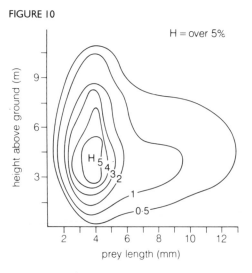

H = over 5%

Two-dimensional niche of the blue grey gnatcatcher (*Poliophila caerulea*) in oak woodlands in California. The contour lines map the percentage of the total diet taken within each part of the niche space. (R. H. Whittaker, S. A. Levin and R. B. Root (1973). Niche, habitat and ecotope. *American Naturalist* **107**, 321–38.)

Fundamental and realised niches

Most animals in isolation can survive a wide range of environmental conditions and can eat a variety of foodstuffs. The range of conditions which a species can tolerate define its fundamental niche. If it is placed in conditions outside its fundamental niche, the animal will die.

There is plenty of evidence that in nature, in the presence of competitors or predators, an animal species only occupies a small portion of its fundamental niche. This is called its 'realised niche'.

A simple illustration of this principle is provided by some observations on two species of ants on the 'Bowling Alley', a narrow strip of grassland amongst woodland at Wytham Woods near Oxford. At various times in the past this site has been used for bowls, illegal cock-fighting, and rearing pheasants. It supports populations of the black ant (*Lasius niger*) and the yellow hill ant (*Lasius flavus*), which often occur together in British pastures.

In the 1950s John Pontin investigated the effect of these two ant species on one another. The black ants nest beneath stones. Each colony of yellow ants builds an ant-hill of fine soil and breeds there. Both species live mainly on aphids and other small insects. Black ants usually search for food above the ground but yellow ants always forage beneath the surface.

Pontin estimated the vigour of each yellow ant colony by the number of new queens which it produced in summer. He put tiles on top of the mounds. The queens congregated beneath them and could be counted.

The yellow ant produced about four times the number of queens when the black ant was absent than when it was present nearby. Furthermore, when black ant colonies were annihilated with sodium fluoride solution, queen production in the adjacent yellow ant colonies markedly increased. These observations suggest that the black ant exerted a considerable influence on the yellow ant. The yellow ants, in the absence of black ones, can occupy the whole of their fundamental and structural niches. In the presence of the black ants the yellow ants were restricted to their realised niches. The result was that they collected less food and produced fewer queens. The two species seemed to be competing for space or a limited food supply.

COMPETITION

Competition occurs when two or more individuals obtain the same resources *which are in short supply*. At least one of the individuals grows or reproduces more slowly than it would have done if its competitors had been absent. In an extreme case, at least one of the individuals dies. The individuals can be members of the same species, when competition is intraspecific, or members of different species, when competition is interspecific.

It is generally accepted that competition between members of the same or closely-related species is usually more intense than that between species which are very different.

Competition is an important process in nature. It has a major influence on the geographical and local distributions of many species. The population sizes of many animals and plants are largely determined and regulated by intraspecific and interspecific competition. Competition between individuals of the same or different species is also a potent evolutionary force.

Since plants and animals obtain rather different resources from the environment, plant and animal competition will be considered separately.

Competition between plants

Individual plants growing next to one another can hardly escape competition, especially if they are all of the same species. In that case they grow at the same time of year, they trap light in a similar way, their roots exploit the same level in the soil, and they require the same proportions of water and nutrient ions. Some plants also compete for pollinators, germination sites, or perhaps even carbon dioxide. Plants hardly ever compete for space. Even in a thick stand of broad bean (*Vicia faba*), only about 1% of the volume above ground is occupied by leaves and 2% of the soil volume is taken up by roots.

Let us analyse the response of broad beans to increasing density (figure 11). As the number of plants per unit area increases seven-fold, the total seed produced only rises by a factor of 1.2. The seed production of individual plants is reduced as they become more crowded, because each stem produces fewer pods. When there are more plants per unit area, each individual obtains a smaller share of the resources, and grows more slowly.

FIGURE 11

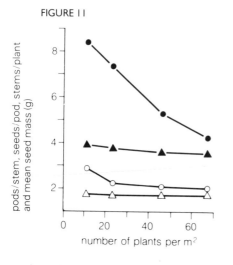

The effect of density on the growth of broad bean (*Vicia faba*). (G. L. Hodgson and G. E. Blackman (1956). *Journal of Experimental Botany*, **7**, 146–65.)

- ● pods/stem
- ▲ seeds/pod
- ○ stems/plant
- △ mean seed mass

pods/stem, seeds/pod, stems/plant and mean seed mass (g)

number of plants per m²

The slower growth rate of the individuals at the higher plant densities is probably the result of two main factors.

(a) When plants are sparse, the sunlight shining on one individual is not intercepted by others, at least in the early stages of growth. In a crowded crop, much of the light which might be shining on one individual is trapped by the upper leaves of other plants. The lower leaves photosynthesise more slowly and contribute less energy to growth than before. They may even respire faster than they photosynthesise. In the humid environment at the base of a broad bean crop, the leaves may be infected by fungi. The result is that at high density, each plant captures less sunlight energy.

(b) The roots deplete the zones around them of water and nutrient ions. The higher the density of broad beans the more interwoven their root systems. The zones of depletion created by one root will reduce the uptake of water and nutrient ions by the adjacent ones. Lack of water will put plants under water stress. The stomata will remain closed for longer, thus reducing the uptake of carbon dioxide in photosynthesis.

Competition for light, water and nutrients reduces the energy available to make new cells. Root growth slows down, reducing the uptake of water and nutrient ions from the soil. Shoot meristems produce fewer leaves and flowers. Individual plants therefore grow more slowly at high than at low densities.

In this experiment all the broad bean seeds germinated at the same time. The plants all grew at about the same rate and obtained the same proportion of the light, water and nutrient ions. Most of them survived until the end of the experiment. This sort of competition, as a result of which plants or animals survive but grow slowly, is called a 'scramble'.

In nature, however, a 'contest' is more usual. If the seeds had germinated at different times, some seedlings would soon have overtopped their neighbours, establishing a dominance which they would never relinquish. In this case the larger plants grow rapidly, obtaining the lion's share of the light, water, and nutrient ions. The smaller plants grow slowly and ultimately die.

So far we have concentrated upon the mechanisms of competition between plants of the same species. Plants of different species compete in a similar way. Yet different species which occupy the same habitat usually differ from one another in form, behaviour and physiology in ways which tend to reduce competition between them. In a plant community, the species differ in growth form, rooting depth, life span, season of maximum growth, pollinators, the heights and shapes of leaves, the possession of toxic compounds, shade tolerance, susceptibility to grazing by a wide range of herbivores and so on. Thus it appears that in nature, intraspecific competition is probably more intense than interspecific competition.

One rather specialised and unusual method by which plants of one species may affect the growth of other species is by the production of toxic compounds. This phenomenon is known as 'allelopathy'. It is well-established that the walnut (*Juglans regia*) produces a compound, juglone, which kills plants growing beneath. In many cases, however, allelopathy is difficult to prove (figure 12).

Competition amongst animals

Animals may compete for food, territories, mates, space, nest sites, overwintering sites or sites safe from predators. There are many examples in this book of competition amongst animals for these scarce resources.

In animals, as in plants, intraspecific competition is probably more intense than interspecific competition. Yet whereas adjacent plants are rooted to the spot, and are forced to compete, most animals can move away from one another and avoid competition. Different animal species in the same community tend to seek cover or nest in different places, eat different sorts or size ranges of food items, catch their food in different ways, or look for food in different microhabitats or at different times of day. In this way interspecific competition is reduced. Closely-related species are designed to occupy different niches. Indeed, the competitive exclusion principle states that no two species can exist together indefinitely in the same community if they occupy the same niche.

FIGURE 12

Zone of bare soil (edge shown by the black line) surrounding a clump of the shrub *Salvia leucophylla* in chaparral at Happy Canyon, California. In the laboratory, compounds released into the air from the leaves of the plant can prevent the germination of 'seeds' of annual grasses. It is also possible that herbivorous rodents live in the *Salvia* clump and eat the vegetation nearby. Can you design an experiment to separate the effects of these two processes? (S. R. J. Woodell.)

THE COMPETITIVE EXCLUSION PRINCIPLE

The idea that 'complete competitors cannot coexist' has stimulated many experiments in which pairs of species have been reared together in the laboratory under constant conditions.

The most famous of these experiments were carried out by Thomas Park on the flour beetles *Tribolium castaneum* and *T. confusum*. They were reared at constant temperatures and relative humidities in vials which contained flour. The flour was renewed every two weeks.

Each species could survive indefinitely on its own in any of the combinations of temperatures and relative humidities which were provided. On the other hand, whenever the two species were reared together, one species always ousted the other. In some cultures it took over a year for one species to become extinct.

At 34°C and 70% relative humidity *Tribolium castaneum* always won (figure 13). At 24°C and 30% relative humidity *T. confusum* won. Under intermediate conditions, the outcome in any particular culture was unpredictable but only one species survived in the end.

If the temperatures and relative humidities within the flour beetle cultures had been altered every twelve hours between hot–moist and cold–dry, to simulate the changes between night and day, the advantage of one species over the other would have been short-lived. The two species would have probably been able to live together indefinitely. In fact many experiments with flour beetles, grain beetles and protozoans have demonstrated that if the niche space is widened two species can co-exist. The species avoid one another to some extent and competition between them is reduced.

The competitive exclusion principle is of dubious value under natural conditions, where the environment is variable in space and time and plenty of niche space is available for different animal species to live together in the

FIGURE 13 Temperature (°C)	Relative humidity (%)	Climate	Percentage of experiments won by	
			T. confusum	*T. castaneum*
34	70	Hot–moist	0	100
34	30	Hot–dry	90	10
29	70	Temperate–moist	14	86
29	30	Temperate–dry	87	13
24	70	Cold–moist	71	29
24	30	Cold–dry	100	0

The outcome of competition between the flour beetles *Tribolium confusum* and *T. castaneum* under different environmental conditions. (Modified from T. Park (1954). Experimental studies of interspecies competition. Temperature, humidity, and competition in two species of *Tribolium*. *Physiological Zoology*, **27**, 177–238.)

same community without competing much with one another. The principle seems impossible to prove with certainty. It seemed possible, for example, that in Britain the North American grey squirrel (*Sciurus carolinenis*) and the native red squirrel (*Sciurus vulgaris*) occupied the same niche. Extra research, however, revealed that the niches of the species differ considerably. Thus one can never be *certain* that two species do occupy the same niche.

SPECIALISTS AND GENERALISTS

Not all species have niches of the same size. Some species, called specialists, are found in a narrow range of environments and feed on a single plant or animal species. They have narrow niches. Other species, known as generalists, are more flexible in their requirements, are more catholic in their tastes, and have broad niches. In temperate regions there are of course many specialists, particularly insects which concentrate upon a single food plant. Nevertheless, it is generally accepted that for many groups of organisms, niches tend to be narrower in the tropics than in temperate regions.

This means that it is possible to pack more niches into a unit area of a tropical than a temperate forest. There are three main reasons for this.

(a) There are many more plant species per unit area in a tropical forest than a temperate one. The reason why so many species evolved is obscure. The diverse flora may be maintained at present by specialist herbivores (p. 59). Whatever the reason for the large number of coexisting plant species in tropical forests, it provides more feeding niches for herbivorous animals.

(b) In a tropical forest, these feeding niches are often available all year round. Flowering and fruiting in closely related species is staggered throughout the year. Specialist fruit-eaters and pollen-eaters can thrive. On the other hand, in temperate regions the production of leaves, flowers and fruits is seasonal. In order to obtain enough food throughout the year, her-

FIGURE 14

An ant 3 cm long, with enlarged mandibles, from the floor of a Peruvian tropical rain forest. In ants, as in many animal groups, more species are known from the tropics than in temperate regions. Amongst tropical insects warning coloration, mimicry and camouflage are developed to a considerable degree. Many startling and grotesque species can be found.

bivorous birds and mammals must be generalists, eating whatever is available.

(c) For both these reasons there are large numbers of specialist plant-eaters in the tropics. This wide range of herbivores can support a wide range of different carnivores and parasites.

Having considered the roles which species are adapted to play in nature, let us now explain how a complex set of factors, including competition with other species, conspire to restrict the members of a species to a particular habitat within a certain geographical area.

3 | *The distribution patterns of species*

The various species are not distributed around the landscape, or water-scape, at random. Each species is restricted to a certain geographical area and to a particular habitat within that area. The habitat preferences of organisms are so precise that if you describe a site to an ecologist (s)he will probably predict accurately the plants and animals which you might find there.

Recently a Dutch botanist, visiting a bog in Ireland for the first time, examined the vegetation. He pointed out that associated with this group of species in Holland he would normally expect to find a cotton sedge, *Eriophorum gracile*. He persisted and ultimately found it. Imagine his surprise when he was told that it had not been found in Ireland before!

In this chapter the reasons why species occur in certain areas and in particular habitats are examined in detail.

THE ROLES OF PHYSICAL, CHEMICAL AND BIOTIC FACTORS

If an animal or plant species survives for a long time in a particular community in a certain geographical region, it must be able to feed and reproduce there. It is well-adapted to the environmental conditions which it encounters. Yet the range of habitats in which it occurs, and its geographical range, may be limited by the effects of other organisms. Its food plant may only occur in one place, or an efficient competitor may prevent it from establishing itself in the community next door. If the food plant spread, or the competitor disappeared, the geographical distribution and habitat range of the species might change.

The effects of other organisms on the distribution patterns of a species are called 'biotic factors'. They include competition, dispersal by humans, and the distribution pattern of the food supply.

If all the biotic factors which restricted the distribution of a species were removed, the species would still be confined to a certain geographical area. This is because it can only tolerate a certain range of physical and chemical conditions. The antarctic fish *Trematomus bernachii*, for example, lives where the water temperature is usually less than 0°C and the sea is covered by ice. It cannot survive above 6°C. Obviously it can never take a holiday in the Bahamas! Apart from temperature, the geographical limits of some species are influenced by relative humidity, oxygen concentration, salinity, water availability and the concentrations of certain ions.

When an ecologist investigates why a certain species has a limited distribution, he often finds it difficult to separate out the most important factor. This is because physical, chemical and biotic factors affect one another so much.

THE FACTOR COMPLEX

Many different environmental factors affect a plant or animal species (figure 15) and they interact and change with time. Consider temperature, for example. The temperature of an organism's environment depends on wind speed, solar radiation, the time of day, the time of year, the aspect and the altitude. Temperature directly affects the rate of water loss from organisms, the rate of cellular respiration and the rate of photosynthesis. It also influences organisms indirectly, by affecting the relative humidity of the atmosphere and the rate at which water evaporates from soil.

FIGURE 15
The factor complex affecting a plant or animal species. All these factors can vary with time. (After Billings.)

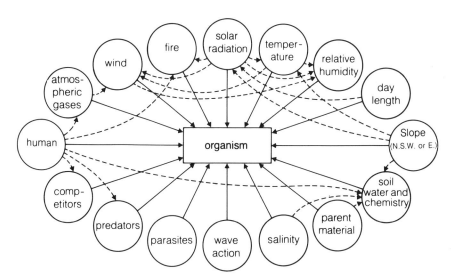

Any organism is therefore affected by a complex of factors in constant flux. In order to illustrate how several factors may simultaneously affect the distribution limits of a species, let us discuss the patterns of two barnacle species investigated by Joseph Connell on rocks on the west coast of Scotland.

Chthamalus stellatus (C for short) and *Balanus balanoides* (B for short) occur on rocks between the tide marks. Species C occurs higher up the rocks than species B (figure 16). Both species are about the same size. They start life as planktonic larvae, which settle out onto rocks as the tide covers them and become clamped to the same spot until they die. They are filter feeders. Their feather-like limbs, which project from the shell, make rhythmical sweeps and strain suspended solids from the water.

Since species C occurs higher up the rocks than species B, it is exposed at low tide for longer periods of time. There is good circumstantial evidence that species C is more tolerant of high temperature and desiccation than species B. This enables it to survive at higher levels.

In fact the upper distribution limits of both species seem to be determined by their tolerances of high temperatures and desiccation. Individuals of both species will die if they are placed at levels above the zones where they normally occur.

FIGURE 16

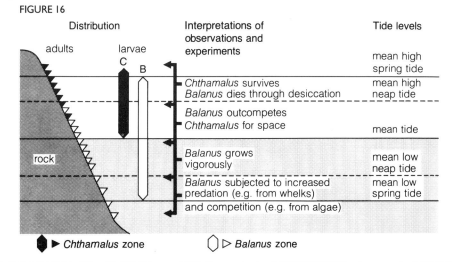

Factors which control the distributions of two barnacle species *Chthamalus stellatus* and *Balanus balanoides*, on a rocky shore between the tide marks, near Millport, Scotland. (After J. H. Connell (1961). The influence of interspecific competition and other factors on the distribution of the barnacle *Chthamalus stellatus*. *Ecology* **42**, 710–23.)

What determines the lower limit of species C? When stones bearing species C were placed much further down the rocks, in the zone dominated by species B, the individuals of species C survived and thrived for a long time. This indicates that species C could certainly withstand long periods under water. In time, however, species B began to colonise these misplaced stones. Its larvae settled in greater population densities and grew faster than those of species C, smothering it, undercutting it and crushing it. The boundary between the two species seems to be determined by competition for space between the two species, with species B winning.

What determines the lower limit of the lower species, species B? Its shorter life span at this level compared with its life span further up the rocks seems to be due to three biotic factors:

(a) increased competition for space within the species,

(b) increased fouling by dense mats of algae attached to the rocks at this level, and

(c) predation by the dog whelk, *Nucella lapillus* (figure 17).

Thus, in the case of the barnacles, the potential limits of their distributions are determined by their tolerances to physical factors like high temperatures and desiccation. Their actual patterns depend largely on biotic factors like competition and predation. In other words, physical factors determine their fundamental niches but biotic factors are important in limiting them to their realised niches (figure 18).

The same is true for other species. The rest of the chapter describes the influence on the geographical and local patterns of species of firstly, physical and chemical factors, secondly, biotic factors, and thirdly, historical factors. At the very end of the chapter the case of the stemless thistle is discussed since it illustrates the influence of all these factors on the distribution of a species.

PHYSICAL AND CHEMICAL FACTORS

The physical and chemical factors which affect the pattern of a species include weather, climate, and all the environmental influences which impinge on a species if it lives in water or in the soil. If the level of any one of these factors exceeds its tolerance limits, the organism will die out.

The main physical and chemical factors in the environment of an organism include salt concentration, wave action, oxygen level, temperature, fire, relative humidity, light intensity, soil chemistry and soil water. These will be discussed in turn.

Salt concentration

In every litre of sea water there are 35 g of dissolved ions, mainly Na^+ and Cl^-. In sea water, ions are over seventy times more concentrated than in

FIGURE 17

Dog whelks attacking barnacles. There are also four limpets (*Patella vulgata*) on the rock.

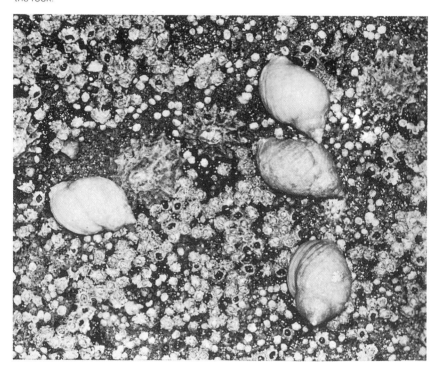

FIGURE 18

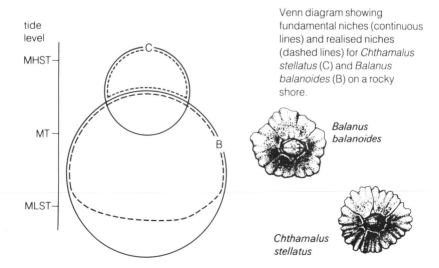

Venn diagram showing fundamental niches (continuous lines) and realised niches (dashed lines) for *Chthamalus stellatus* (C) and *Balanus balanoides* (B) on a rocky shore.

Balanus balanoides

Chthamalus stellatus

tide level

MHST

MT

MLST

fresh water. It therefore exerts a low solute potential (-24 atm $= -23.7$ bars $= -2370$ kPa). Only those organisms with well-developed mechanisms of salt regulation can live in it. One example from each end of the salinity spectrum will suffice. The spider crab *Maia squinado* is marine and dies if the salt water in which it lives is only slightly diluted or concentrated. The crayfish (*Astacus* spp.), on the other hand, lives in freshwater. Although it can increase the salt content of its blood slightly in response to a greater concentration of salt in the water, it dies even in half strength sea water. The reasons why these organisms die in conditions to which they are unaccustomed are obscure, but they probably cannot regulate the water content of their body fluids.

The geographical distributions of plants are also affected by salinity. Some higher plants which are confined to salt marshes around the coasts require a salty soil to survive, like the glasswort (*Salicornia* spp.) Salt marshes maintain a distinctive set of species. Most plants which grow inland, on non-saline soils, cannot tolerate the high salt concentration in estuarine mud.

Wave action
Wave action particularly affects animals and plants which live between the tide marks. Those organisms which cannot withstand buffeting by waves or the scouring action of the sea are confined to sheltered sites, like salt marshes. The animals and plants which can live on exposed rocks are attached firmly and protected from the force of the waves. Barnacles, for example, clamp themselves firmly to stone and have an exoskeleton which is impregnated with calcium carbonate. Many seaweeds, like the kelps (*Laminaria* spp.) have tough branched holdfasts inserted into clefts in the rock, and their thalli are flexible yet strong.

Oxygen level
The oxygen level in the environment is important to organisms because they need oxygen for cellular respiration. In a litre of air there are 210 cm^3 of oxygen. In a litre of water saturated with oxygen, however, there is always less than 10 cm^3 of the gas. Moreover, the rate of diffusion of oxygen in water is far slower than its rate of diffusion in air. Aquatic animals and plants need to extract oxygen from the water. Thus any reduction in the oxygen level in water may be critical for animals or plants living in rivers, streams, lakes, ponds, estuaries or the sea.

The concentration of dissolved oxygen is high in running waters like rivers and streams. Stonefly nymphs are found only in well-aerated streams. In Britain, nutrient-poor lakes like mountain tarns on granite have a high oxygen content in summer and are inhabited by fish like brown trout (*Salmo trutta*), whitefish (*Coregonus* spp.) and char (*Salvelinus alpinus*).

In contrast, nutrient-rich lakes have a low dissolved oxygen content because of the oxygen taken up from the water by bacteria which are decomposing dead algae. They support members of the carp family like roach (*Rutilus rutilus*) and bream (*Abramis brama*), and have leeches, water lice (*Asellus aquaticus*) and tubifex worms living on the bottom. These organisms do not necessarily prefer deoxygenated water. They may be excluded from oxygenated conditions by competition.

The effect of a marked change in the dissolved oxygen content of water was demonstrated in the cleaning up of the River Thames in London between 1950 and 1978. In 1950 there was no dissolved oxygen. The bacteria, in decomposing the sewage effluents, had removed the oxygen from the water. There were no fish to be seen. Thirty-two years later, the concentration of dissolved oxygen is around 90% of the maximum possible for much of the year and over a hundred species of fish have been found in the river, including salmon (*Salmo* spp.).

Temperature

Temperature can affect the distribution patterns of organisms. Many plant and animal species have phases in their life-cycles, like seed germination or the development of the young, which require particular temperature conditions for success. Despite this, good correlations between the distribution limits of organisms and isotherms of temperature are rare.

One correlation with temperature which is fairly well established involves the mosquito *Aedes aegypti*. Its distribution boundary closely follows the 10°C winter isotherm (figure 19). The lower lethal temperature for adults, larvae and eggs is about 10°C.

Fire

Fires caused naturally by lightning are frequent in many habitats with a dry season. In general, they maintain a shrubby or grassy vegetation, at the expense of the long-lived trees which would otherwise make up the climax vegetation. Fire is particularly significant in African savannah, Mexican and Californian chaparral, and some eucalyptus forests in Australia.

Relative humidity

Like fire, relative humidity affects only terrestrial animals and plants. The relative humidity of the air is the mass of water vapour actually present, expressed as a percentage of the mass of water vapour held by saturated air at the same temperature. The lower the relative humidity, the faster water evaporates from cells exposed to the air.

Since relative humidity varies in both space and time, it influences the local distributions of many mosses, liverworts, algae, woodlice, slugs and snails. In these organisms large areas of moist tissue are exposed to the atmosphere.

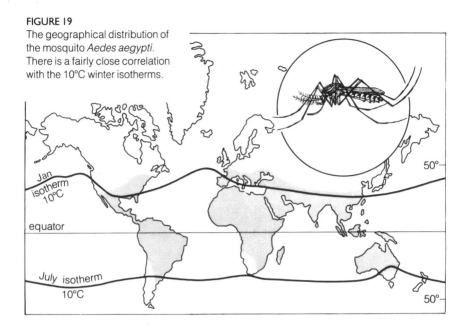

FIGURE 19
The geographical distribution of the mosquito *Aedes aegypti*. There is a fairly close correlation with the 10°C winter isotherms.

Jan isotherm 10°C

equator

July isotherm 10°C

50°

50°

Woodlice, for example, occur mainly beneath bark, under stones and beneath leaf litter, where the relative humidity is particularly high. They can easily lose water vapour through their cuticles and gas-exchange organs. Woodlice find and remain in such moist spots because of a kinesis, an instinctive behavioural mechanism which involves the movement of the whole organism. In fact most animals have a built in concept of the micro-habitat which they should occupy and can select it instinctively. Like woodlice, most terrestrial invertebrates can probably sense the relative humidities of their environments and react in ways which increase their chances of survival.

Light intensity
Light intensity certainly influences the geographical and local distributions of plant species. The intensity of solar radiation reaching the earth's surface at a particular place depends upon the latitude, the season of the year, the time of day and the extent of cloud cover. There is considerable variation in light intensity on a geographical scale. A site on the equator, for example, receives three times as much solar radiation throughout the year as an equivalent area at 50° latitude. This variation in light intensity produces geographical differences in temperature and evaporation rate. It is therefore impossible to be sure that the limits of a species are determined by light intensity as such.

The soil

Soil is the medium in which plants grow. A fertile soil has six main constituents: mineral particles (derived from rock), humus, water, nutrient ions, air and organisms. Soil factors, sometimes called edaphic factors, vary considerably on both a local and a geographical scale and affect the patterns of plant species and soil animals.

The bulk of most soils consists of mineral particles. These are formed from bedrock by several processes which include:

(a) erosion of rock by moving water, which results in the formation of alluvial soils of silt and clay in river deltas and flood plains,

(b) erosion by glaciers,

(c) blasting of rock by wind-blown sand,

(d) the differential expansion and contraction of different layers of rock, as they are heated by the sun during the day and cool down at night,

(e) the freezing and expansion of water in cracks in the rock,

(f) the breaking up of rock by the roots and root exudations of plants, called 'biological weathering'.

The mineral particles ultimately produced by these processes differ in size and chemical composition. At one extreme the large coarse sand particles (2.0–0.2 mm diameter) are composed of silicon dioxide, a giant molecule. The large spaces between the particles make sandy soils well-drained and well aerated. Nutrient ions are often sparse in sandy soils because the uncharged sand particles do not attract them and drainage is rapid. At the other extreme, clay particles (<2 μm in diameter) are composed of potassium aluminium silicates. The tiny gaps between the particles hold water by surface tension; clay soils tend to be waterlogged and poorly aerated. They are often rich in nutrients however, because positively-charged ions, held by negative charges on the clay, are sandwiched between the flat clay plates. In practice most soils have particles of a variety of sizes and are known as loams (figure 20).

The mineral particles are surrounded by humus. This is a sticky black mixture of organic compounds derived from the decay of detritus, which consisted of dead parts of organisms and their excreta. Humus is undergoing continual breakdown, mainly by bacteria and fungi. This is important to the nitrogen cycle in the soil; decay of the organic nitrogen compounds locked up in the humus releases the ammonium ions which ultimately provide nitrate, the main nitrogen source for plants (p. 117). Humus holds water, and its negative charges hold positively-charged nutrient ions. It is food for earthworms, whose activities markedly influence soil (p. 32). In addition, humus glues together mineral particles into larger structures known as 'crumbs'.

Typical crumbs (figure 20) are 3–5 mm in diameter. Soils with a crumb

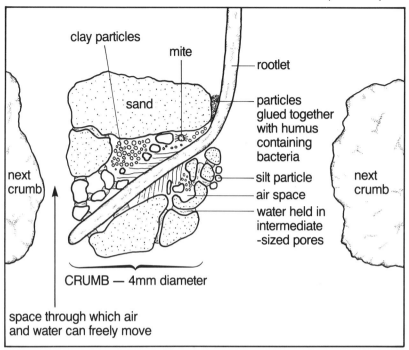

clay particles

mite

sand

rootlet

particles
glued together
with humus
containing
bacteria

next
crumb

silt particle

air space

water held in
intermediate
-sized pores

next
crumb

CRUMB — 4mm diameter

space through which air
and water can freely move

FIGURE 20
A section through some soil crumbs in a typical loam.

structure tend to be well-drained and aerated, since there are spaces between the crumbs through which water can trickle and air can pass. At the
same time water and nutrients, held within the crumbs, are available to
roots.

The spaces within and between crumbs are filled with air and water.
The water, known as the soil solution, is held between the mineral particles
and in humus in gaps which differ in size. The water in the largest pores is
evaporated, and taken up by roots, most easily. In the smallest pores,
particularly those between the clay particles, the water is held firmly by surface tension and only the most thirsty rootlets can extract it. If a soil dries
out by evaporation or drainage, the water which remains becomes confined
to smaller and smaller pores.

The base of many soils is saturated with water in a zone known as the
water table. The water table may be continuous with the water in rivers and
streams nearby. In some soils water lost by evaporation and drainage is
replaced by capillary rise from the water table. Water is also replaced by
rain, snow, fog and dew.

Dissolved in the soil water are the ions which are essential for plant
growth. Their concentration depends on the chemical constitution of the
rock from which the soil was formed, and the activities of bacteria and other

organisms. As ions are absorbed by roots, they are replenished from three main sources:
(a) from the ions attached to the negatively-charged surfaces of clay particles and humus,
(b) from the decay of humus, parts of dead organisms, and excreta,
(c) from rainfall.

The more soil water there is, the less air. The oxygen in the soil atmosphere is used in aerobic respiration by the organisms which occupy the spaces between the soil particles. These include earthworms, nematode worms, mites, ants, insect larvae and pupae, bacteria, fungi and roots.

Earthworms contribute to soil fertility. They remove detritus from the soil surface, and after digesting it, release nutrient ions which roots can take up. Worm channels improve drainage, help aeration and ease root growth. The 'casts' of soil which worms egest create crumbs. Worms counteract leaching, the washing of nutrients from the upper layers of the soil, by eating soil from the lower layers and depositing it on the surface. Several experiments have shown that crops grow better in the presence of earthworms.

Vertical sections of the soil are known as soil profiles. The arrangement, composition and thickness of the various layers ('horizons') which occur with increasing depth provide a record of the history of the soil. In acidic soils with a pH 5.5, for example, the soil is often podsolised (leached and 'ash-like'). Earthworms do not occur in acidic soils and so the layers hardly mix. The detritus only breaks down slowly and accumulates on the surface (figure 21). Below this the upper layer of mineral soil becomes bleached as the aluminium and iron (III) salts are washed down the profile, attached to organic acids derived from detritus. These salts accumulate in an 'iron pan', a brown layer which often sets hard and inhibits drainage and root growth. The relevance of soil profiles to plant distribution patterns is obscure. Variations in soil chemistry, however, have a considerable influence on the patterns of many plants and soil animals.

Soil chemistry

Firstly, many plants only occur within a limited range of soil pH. Secondly, the nutrient levels required for optimum growth differ from one species to another. Thirdly, high levels of certain ions can be toxic to some plants.

Some plants, known as calcicoles, are characteristic of soils with pH >7 (e.g. stemless thistle, *Cirsium acaule*, p. 42) and others occur only on acidic soils (e.g. ling, *Calluna vulgaris*). Research suggests that pH can affect plant distribution by affecting the solubility of ions. Iron(III), for example, may be unavailable to the roots of calcifuges in alkaline soils because it is deposited on the soil particles as iron(III) hydroxide. This may

FIGURE 21

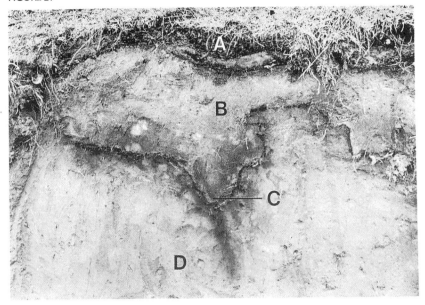

The 'profile' of an iron-pan podsol developed over sandstone in the North York Moors. The vegetation was ling (*Calluna vulgaris*) and conifers. No earthworms were present. Several 'horizons' (horizontal layers) are visible: **A** is accumulated humus, under conifer leaf litter, **B** has been bleached because salts have been washed downwards, **C** is the iron pan, and **D** is sandy soil leading down to bedrock. (J. R. Etherington, 1978).

be why such plants exhibit the symptoms of iron deficiency on alkaline soils and why the symptoms can often be relieved by spraying the leaves with a solution (Fe–EDTA) which supplies the missing iron.

To illustrate that certain species need high levels of certain nutrient ions for maximum growth, let us consider some experiments on nettles. Nettles (*Urtica dioica*) are precise indicators of old habitations, dungheaps, seabird colonies and rabbit warrens. Donald Pigott and Ken Taylor, investigating the distribution of nettles in some Derbyshire woodlands, suspected that nettles might indicate high levels of soil nitrogen and phosphorus.

When they analysed nettle shoots, nettles contained far more nitrogen and phosphorus per gram than two other woodland herbs, dog's mercury (*Mercurialis perennis*) and tufted hair grass (*Deschampsia caespitosa*). Nitrogen levels, however, were similar in soils beneath all three species. Nettle seedlings showed little response when they were grown in areas of dog's mercury and tufted hair grass, and supplied with extra nitrogen in the form of ammonium nitrate (figure 21).

When calcium phosphate was added to the soil in areas in which nettles did not normally grow, nettle seedlings planted in the soil responded dramatically provided that the light intensity was high. In fact the growth rate of nettles is directly proportional to the phosphate concentration in the soil. The best way to compare the levels of phosphate ions available to plant roots in soils is to grow nettles from seed on the soils and to compare their growth rates.

FIGURE 22

Soil from	No added nutrients	+ Ca(H_2PO_4)_2	+ NH_4NO_3	+Ca(H_2PO_4)_2 + NH_4NO_3	+ All essential nutrients except P
Buff Wood, Cambs.	6	92	5	125	5
Cressbrookdale, Derbys.	1	248	1	281	1
	(many dead)		(many dead)		(many dead)

Mean dry mass (mg) of nettle plants grown from seed. The nettles were grown in a glasshouse on woodland soil collected from places where dog's mercury was growing and nettles were absent. (From C. D. Piggott and K. Taylor (1964). The distribution of some woodland herbs in relation to the supply of nitrogen and phosphorus in the soil. *Journal of Ecology* **52** (Suppl.), 175–85.)

Thus it seems that at least on open woodland soils the availability of phosphate in the soil may be a major factor limiting the distribution of nettles. If this is true elsewhere it explains why nettles are good indicators of the past activity of humans and other animals. Animal egesta contains relatively high concentrations of phosphate. Unlike the nitrate ion, which is rapidly washed into groundwater by rainfall, the phosphate ions (PO_4^{3-}, HPO_4^{2-}, $H_2PO_4^-$) rapidly form insoluble phosphates which may persist for thousands of years.

Nettles grow very rapidly under nutrient-rich conditions, cast a dense shade and grow vigorously by undergound rhizomes. Under these conditions they are capable of outcompeting other species.

Some rock outcrops bear distinctive vegetation. Soils developed over serpentine rock have high levels of nickel and chromium and low levels of calcium and phosphate in the soil solution. Their vegetation is usually quite different from that of their surroundings (figure 23) and they only support a sparse cover of plants. Serpentine sites often look remarkably similar to the spoil heaps of spent ores created by man, which contain high levels of aluminium, copper, lead or zinc.

Yet one plant's poison is another plant's food. A few species, like the herb *Bechium homblei* in Zimbabwe, can accumulate high concentrations of metal ions in their cells and often occur at sites with soils rich in metals.

FIGURE 23

The effect of rock type on vegetation, in the San Rafael mountains, California. The boundary between limestone, on the left, and serpentine rock, on the right extends down the centre. Compared with the limestone soil, the serpentine soil is rich in nickel, chromium and magnesium, and poor in calcium. The serpentine site supports open dwarfed vegetation of different species composition to the *Pinus sabiniana* forest on the left. (S. R. J. Woodell.)

B. homblei can tolerate soil copper concentations up to 15 000 ppm and soil nickel of nearly 5000 ppm. Man makes use of this in an unexpected way. These plants are used by geologists in mineral prospecting, to pinpoint the sites which might be mined to yield valuable metals.

Soil water

Soil water levels have an obvious and dramatic effect upon the distribution of plants. A common sight is a dry grassy field, with rushes and sedges in the hollows, and a completely different set of plants growing beside the edge of a stream.

Soil water may exert its influence on plant distribution in subtle ways. Let us discuss two examples.

(a) In some Cambridgeshire woodlands, the herb dog's mercury is confined to the dry slopes and does not occur in the wet areas. In wet soils, oxygen diffuses to roots more slowly and iron(III) is reduced to iron(II). The lower parts of the soil profile become blue–green (iron(II) salts) speckled with black (manganese dioxide), and the soil is known as a 'gley'. On this basis Michel Martin investigated why dog's mercury did not occur in the wetter areas.

When blocks of soil with dog's mercury 'turf' growing on them were transplanted to a wetter site, the leaves of the plants began to yellow during the following spring. Root growth had been seriously slowed.

In the laboratory the roots of dog's mercury seedlings were grown in enclosed water cultures aerated with different gas mixtures. Low oxygen concentrations did not severely limit root growth. When seedlings were grown in various concentrations of manganese sulphate, the shoots died at high concentrations but the roots continued to grow. The roots died, however, when seedlings were grown in iron(II) sulphate solutions with more than 4 ppm Fe(II). These results suggest that high levels of toxic Fe(II) ions in the soil in spring probably prevent dog's mercury from growing in the moister places.

(b) Many plants which in nature are restricted to waterlogged soil are capable of growing on drier soils in the absence of competition. They are fugitives. In nature they are outcompeted on drier soils, but can survive on wet soils from which their competitors are excluded.

The water ragwort (*Senecio aquaticus*), for example, survives on wet soils by being an alcoholic. Its root cells can withstand much higher concentrations of ethanol, formed by anaerobic respiration, than those of closely-related species which are normally restricted to dry soils.

BIOTIC FACTORS

Most species only occupy a fraction of the geographical area in which their tolerance of physical and chemical factors equips them to live. This is because other species affect their distributions. These biotic factors are of several types.

Firstly, the ranges of many species are limited by competiton with other species. Competition has already been dealt with in chapter 2.

Secondly, organisms live near their food. Many herbivores occur only near their food plants. Most predators are limited to places where their prey is abundant. Parasites and mutualists are restricted to the habitats of their hosts. Food plants, predators, parasites and mutualists will be mentioned in turn.

Food plants

Insect herbivores often specialise on a single food plant. Most plant species support characteristic herbivores.

The insects seem to be attracted to their food plants by chemicals called phagostimulants. Species of the plant St. John's Wort (*Hypericum* spp.) for example, contain a compound called hypericin. This protects them from attack by most insects. Yet hypericin seems to be a phagostimulant for beetles of the genus *Chrysolina*. They attack the St. John's Wort with relish, especially the flower heads and the upper leaves, which contain hypericin in the highest concentrations.

This sort of interaction between food plant and herbivore might have two consequences. In the first place, many specialised plant feeders only live in places where their food plants occur. In the second place, by eating the food plant, the herbivore might restrict the plant's distribution. Both these statements are true for *Chrysolina* and St. John's Wort.

The beetle *Epitrix atropae* is another example of a case where the distribution of an insect depends on the presence of its food plant. This insect feeds only on deadly nightshade (*Atropa belladonna*). At Wytham Woods, near Oxford, rabbits created bare patches in the grassland on which the seeds of deadly nightshade could germinate, and both the plant and its beetle were frequent. After the rabbit population was eliminated by the myxomatosis virus in 1954–5, the deadly nightshade rapidly declined, and with it the beetle, which has not been seen at Wytham since the mid 1950s.

Predators
Some secondary consumers are restricted in their distributions by the patterns of their prey. Predators tend to congregate at sites where their prey is abundant. For example, blue whales feed mainly on krill, which are crustaceans, <12 cm long which they filter out of the sea. At one time blue whales probably ate about 150 Tg year^{-1} of krill, which feed directly on plant plankton. The Discovery expedition found in 1926–7 that blue whales had congregated in areas of massive local concentrations of krill in the Arctic seas.

Parasitism
Parasites usually specialise on a single host, or a narrow range of host species. Parasites occur where their hosts occur.

Black stem rust of wheat (*Puccinia graminis*), for example, an important fungal pest in Britain, has barberry (*Berberis vulgaris*) and wheat (*Triticum aestivum*) as alternative hosts. Farmers frequently eliminate barberry from hedgerows to reduce the chances of an outbreak of the disease.

The case of the alder wood wasp (*Xiphydria camelus*) and its insect enemies illustrates the host specificity and the complexities of parasitism in nature. The female wood wasp, after mating, uses her ovipositor to drill a hole in the bark of a dead alder stump and lays an egg at the base of the hole. This egg, or the resulting larva or pupa, can be attacked by one or more of the four host-specific parasites of the wasp (figure 25). Their life cycles depend entirely on the dead alder wood with the alder wood wasp living inside it.

Mutualism
Mutualism occurs when two or more species habitually occur together in a close association from which both species benefit. In many cases of mutual-

FIGURE 24

A male great diving beetle (*Dytiscus marginalis*) parasitised by larvae of the red water mite *Hydrachna globosa*. These parasitic larvae need a large insect host like a water scorpion (*Nepa cinerea*) or a water beetle. Although the adult mites are free-living carnivores, their geographical distribution may be restricted to water bodies in which suitable hosts occur for the larvae.

FIGURE 25

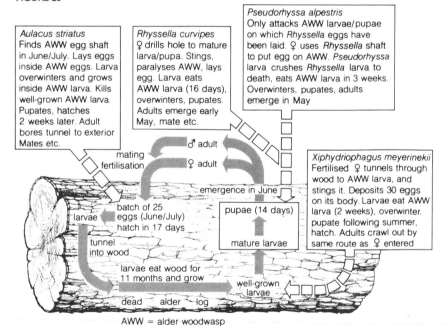

Aulacus striatus
Finds AWW egg shaft in June/July. Lays eggs inside AWW eggs. Larva overwinters and grows inside AWW larva. Kills well-grown AWW larva. Pupates, hatches 2 weeks later. Adult bores tunnel to exterior Mates etc.

Rhyssella curvipes
♀ drills hole to mature larva/pupa. Stings, paralyses AWW, lays egg. Larva eats AWW larva (16 days), overwinters, pupates. Adults emerge early May, mate etc.

Pseudorhyssa alpestris
Only attacks AWW larvae/pupae on which *Rhyssella* eggs have been laid. ♀ uses *Rhyssella* shaft to put egg on AWW. *Pseudorhyssa* larva crushes *Rhyssella* larva to death, eats AWW larva in 3 weeks. Overwinters, pupates, adults emerge in May

Xiphydriophagus meyerinekii
Fertilised ♀ tunnels through wood to AWW larva, and stings it. Deposits 30 eggs on its body. Larvae eat AWW larva (2 weeks), overwinter, pupate following summer, hatch. Adults crawl out by same route as ♀ entered

mating
fertilisation
♂ adult
♀ adult
emergence in June
pupae (14 days)
mature larvae
batch of 25 eggs (June/July) hatch in 17 days
larvae
tunnel into wood
larvae eat wood for 11 months and grow
well-grown larvae
dead alder log

AWW = alder woodwasp

The life histories of the alder wood wasp (*Xiphydria camelus*) and the four species of parasitic insect which depend on it for food. Notice how the life cycles of the parasites are synchronised with that of their host. (After the film 'The Alder Wood Wasp and its Insect Enemies' Oxford Scientific Films.)

ism, the geographical distribution of the smaller of the two species is determined by that of the larger partner. On the other hand, the presence of the smaller species increases the range of habitats which the larger partner can occupy.

As an example, consider the mycorrhizal association between some fungi and the roots of higher plants. Many mushrooms and toadstools commonly found in beech (*Fagus sylvatica*) woodland, like the death cap *Amanita phalloides*, are the fruiting bodies of fungi which form mycorrhizae with beech roots and require beech to survive. The fungus obtains its organic compounds from the tree. In return, it provides the roots with ions like nitrates and phosphates absorbed from the soil. This allows beech to thrive even on soils of low nutrient status.

Other examples of mutualism include:

(a) The presence of the green alga, *Chlorella*, in the cells of the green hydra, *Chlorohydra viridis*. The alga obtains carbon dioxide from the hydra cells, photosynthesises, and donates carbon compounds to its partner.

(b) The occurrence, in the guts of termites and ruminants, of bacteria and protozoa which can digest cellulose and lignin and convert them into compounds useful to the host animal.

(c) The relationship in warm tropical seas between cleaning shrimps or fish, like the cleaner wrasse (*Labroides dimidiatus*), and larger fish which queue up to be cleaned (figure 26). When a larger fish approaches, the conspicuously striped wrasse begins a complex dance which signals to the larger fish that the wrasse is a cleaner. The large fish stops dead with gill covers raised, and allows the wrasse to clean its body and gills of parasites. The wrasse gains a source of food and the larger fish lives a longer and more vigo-

FIGURE 26

The cleaner wrasse *Labroides dimidiatus*, small and striped, cleaning a recticulate damselfish (*Dascyllus reticulatus*).

rous life. The value of the association to the larger 'client' fish was illustrated when all the 'cleaners' were removed from a large area of reef off the Californian coast. After two weeks many large fish had disappeared and those which remained had numerous parasites and festering sores.

(d) The association, studied by Dan Janzen in Central America, between swollen-thorn *Acacia* trees and ants of the genus *Pseudomyrmex*. The fertilised queen ants establish colonies within the hollow spines of the *Acacia*. The larvae are fed on protein-rich 'Beltian bodies' which protrude from the edges of the leaves, and the adults suck nectar from nectaries on the petioles. These sources of food are available all year round.

In return, the ants may attack caterpillars. When Janzen cut down some *Acacias* so that they suckered from the base, the suckers colonised by ants remained healthy but those without ants were rapidly devasted by herbivorous insects. The ants may also sever the stems of vines which grow nearby. When ant colonies on mature *Acacias* were removed with insecticide, the trees were smothered within months by creeping vines.

HISTORICAL FACTORS

The present geographical distributions of many species cannot merely be explained in terms of physical, chemical and biotic factors. Firstly, the present range of a species may be limited by inefficient dispersal in the past. Secondly, humans have dispersed many species over wide geographical areas. Thirdly, some species were once widespread but have since been eliminated over much of their range by climatic change or human activity. These factors will be considered in turn.

Dispersal

The geographical distributions of many species are considerably influenced by dispersal, or lack of dispersal. The efficiency with which some animal and plant species are dispersed is illustrated by the re-invasion of the island Krakatau in the East Indies after its massive explosion on 26 August 1883. Ten cubic kilometres of rock were blown away from the volcano, leaving a lifeless surface, 40 km away from the nearest island. A spider was seen after nine months and after three years there were cyanobacteria, eleven species of fern and fifteen flowering plant species on the island. Coconut trees had arrived after ten years, and twenty-five years after the explosion there was dense forest and at least 263 animal species. After fifty years, the animal life included forty-seven species of vertebrates: 36 birds, 5 lizards, 3 bats, a rat, a crocodile and a python!

The most important dispersal agents for animals and plants are air, water and animals.

(a) Air. Many plant species are efficiently dispersed by wind. Some have tiny spores or seeds, and others produce seeds or fruits with plumes or wings. Nets attached to masts or kites catch a bewildering variety of animals, especially aphids, small flies, spiders, gnats and parasitic wasps. These weak fliers are probably caught on upward convection currents and blown for hundreds of kilometres.

Flight is also a means of active aerial dispersal for insects and birds.

(b) Water. The fruits and seeds of plants are often carried by water, like the seeds of the water lily and fruits of the coconut.

Many animals incapable of flight, like the lizards and the rat which reached Krakatau, may have reached offshore islands by clinging to tree trunks or pieces of driftwood.

(c) Animals. Dispersal of plants by animals is usually short-range. Fruits, seeds, spores and eggs are often carried on the outsides of organisms without being eaten. Some seeds, such as those of the violets (*Viola* spp.) possess oil bodies which attract ants. The ants then transport the seeds. Birds and mammalian herbivores may eat seeds and fruits and void them unharmed far away. Many birds can migrate long distances in a short time and carry the seeds with them.

Dispersal by humans
Deliberate and accidental introductions of species to continents and islands have been made by humans at an increasing rate. Some examples of these are mentioned in figure 27.

Introduced species have sometimes caused such economic damage (see pp. 70–2) that customs authorities spend a great deal of money on the inspection and quarantine of imported animals, crops and timber. The success of many introduced species indicates that the major factor which previously limited their geographical ranges was a dispersal barrier. Dispersal barriers are easily overcome by the use of human modes of transport such as aeroplanes and ships.

Land-use history
In the lowland agricultural landscape it is not surprising that land-use history has markedly influenced the local patterns of plants and animals. Did the parish make its living by forestry, or by grazing? Was the land reclaimed from the sea? How long ago was a field last ploughed or burnt? Has it been treated with fertilisers or herbicides?

One example will illustrate the importance of factors like these.

Woodlands which have never been cultivated and grasslands which have never been ploughed often support groups of uncommon species. This suggests that they have retained some of their original species whilst cultivation has removed them from the surrounding countryside. The wild service tree (*Sorbus torminalis*) and the Midland hawthorn (*Crataegus*

FIGURE 27
Some famous introductions by humans of species to new areas of the world.

Species	Introduction	Spread
European starling (*Sturnus vulgaris*)	Europe to New York 1890.	Now present in all U.S.A. and most of Canada.
Rabbit (*Oryctolagus cuniculus*)	Europe to Britain in late 12th century.	Originally bred for fur and meat. Escaped. Wild population expanded in 18th and 19th centuries. Major farm pest until myxomatosis in 1953–5. Only 1% survived. Now expanding again.
Coypu (*Myocaster coypus*)	Fur farms established in Britain 1929.	Now confined to 6000 km^2 of East Anglia. Trapping and cold weather reduced population from 200 000 (1962) to 3000 (1978) to virtual extinction (1989).
Prickly pear cactus (*Opuntia* spp.)	South America to Australia 1900.	Covered 24 000 000 ha of former grazing land by 1925. Eggs of moth (*Cactoblastis*) imported from Argentina 1925. Moth achieved rapid control.
Oxford ragwort (*Senecio squalidus*)	Southern Italy to Oxford Botanic Garden (1794).	Spread to London along railway ballast and then over much of Britain.
Canadian pondweed (*Elodea canadensis*)	Canada to British Isles 1836.	Spread along canals and waterways and blocked many in late 19th century. Has since declined.
Chestnut blight (*Cryphonectria parasitica*)	Asia to New York 1900.	Has almost eliminated the American Chestnut (*Castanea dentata*) from deciduous forests in E. United States, in which it formerly made up 40% of the large trees.

oxycanthoides) indicate old woodlands. Old chalk grasslands which have not been ploughed in the last 200 years may contain species like the Pasque flower (*Pulsatilla vulgaris*).

We have dealt with the influences of physical, chemical, biotic and historical factors on the geographical and local patterns of different species. But in practice these individual factors interact in a complex way. The rest of this chapter is devoted to an investigation by Donald Pigott of the way in which these factors interact to influence the distribution of a single species, the stemless thistle.

THE STEMLESS THISTLE (*CIRSIUM ACAULE*)

The stemless thistle is characteristic of chalk and limestone grasslands in Britain and Europe (see figure 28). It is a perennial belonging to the daisy family (*Compositae*), with spiny leaves 10–20 cm long which are pressed to the ground. The leaves are arranged in rosettes, which are groups of leaves

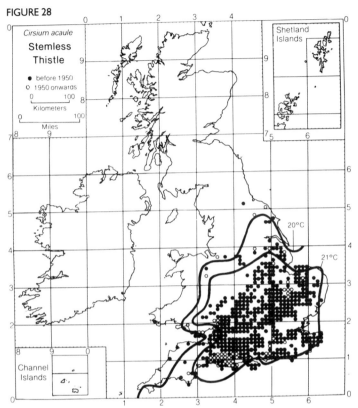

FIGURE 28

The distribution of *Cirsium acaule* in the British Isles. Each dot represents at least one record in a 10-km square of the National Grid. The lines show average means of daily maximum temperatures in August. (From *The Atlas of British Flora*.)

radiating from the central point where they join the rootstock. It has a horizontal branched underground stem which turns up at the ends to produce rosettes. This provides a method of vegetative reproduction. Rosettes growing near one another are often products of the same underground rhizome.

In summer, some of the rosettes flower. Each flowering rosette produces 1–4 flowering heads, each called a capitulum, on very short stems. Each of the 50–150 florets in a capitulum is capable of forming a single viable fruit.

We can ask several questions about the distribution of this species. Firstly, why is it confined to grasslands? Secondly, why is it usually found on soils which are high in calcium carbonate? Thirdly, why does it abruptly reach the northern limit of its distribution in the Northern Pennines? These questions will be discussed in turn.

FIGURE 29

Sheep-grazed chalk grassland, a typical habitat for the stemless thistle. Aston Rowant National Nature Reserve, Oxfordshire (1976). There were no shrubs on the site in the late 1940s and early 1950s. The elimination of the rabbit by myxomatosis in 1953–5 temporarily reduced the grazing pressure and allowed the shrubs to become established. The mounds have been built by the yellow hill ant, *Lasius flavus*, over the last sixty years. The stemless thistle does not grow on them.

Why is the thistle confined to grasslands?

The stemless thistle grows most vigorously where the grasses have been grazed down to a height of less than 15 cm by sheep, rabbits or cattle (figure 29).

If grazing is relaxed, the rosettes of the thistle may persist for many years but they ultimately die. New seedlings cannot become established because they are out-competed by tall grasses. The inability of the thistle to withstand shade may prevent it from growing in scrub and woodland.

Pigott tested the effect of shade on newly germinated seedlings which were sown under woodland canopies that only transmitted 1–5% of the light in the photosynthetic wavelengths. They rapidly died.

Why is the thistle absent from acid soils?

The stemless thistle occurs mainly on soils 15–45 cm deep in which the pH is usually above 5.5 and there is some calcium carbonate within rooting depth. Why is the species absent from soils of pH 4–5?

Seedlings of the thistle are never found in nature on such acid soils, which suggest that the plants are eliminated before the seedling stage. When Pigott sowed the fruits on to acid soils, the seeds germinated, but the radicles became discoloured and stunted. The plants rapidly died.

The stunted root growth is symptomatic of aluminium toxicity. Together with a purple coloration beneath the leaves which is characteristic of phosphorus deficiency, it is one of the symptoms which often appear in the seedlings of chalk-loving plants (known as calcicoles) when grown in acid soils. There is plenty of evidence that calcicoles cannot survive even a low concentration of aluminium ions in the soil solution. The sedge *Carex lepidocarpa*, for example, occurs only in those bogs which have <1 ppm ($<3.7 \times 10^{-5}$ moles 1^{-1}) of aluminium ions in the water.

The concentration of aluminium ions in the soil solution is related to pH in the following way. At high pH, aluminium ions leave the soil solution. They are deposited as insoluble aluminium hydroxide on the surfaces of the mineral particles. As the pH declines, the aluminium ions dissolve more and more in the soil solution, and come into contact with plant roots. They may form insoluble aluminium phosphate within the plant, thus preventing efficient phosphate uptake from soils which are already low in phosphate.

Stemless thistle seedlings were grown in solutions containing 3×10^{-4} moles 1^{-1} of aluminium salts. Their radicles became discoloured and stunted, just as they did on acid soils. This is circumstantial evidence that it is the high concentration of aluminium ions in the soil solution which prevents the thistle from growing on acid soils.

When Pigott analysed the leaves of the stemless thistle, the concentrations of nitrogen and phosphorus were so low that they would be associated with deficiency in many other species. On more fertile soils, the thistle is outcompeted by grasses which can respond to fertile conditions by growing rapidly.

The stemless thistle therefore requires a soil which (a) has a pH>5.5, (b) has a low concentration of aluminium ions in the soil solution, and (c) is poor in nitrogen and in phosphate ions. These edaphic factors tend to restrict the species to unfertilised pastures on chalk or limestone.

Climate and fruit production

The north-western limit of the stemless thistle in Britain seems to be determined by summer warmth, because its boundary follows very closely the 21°C in July and 20°C in August summer isotherms. Along this boundary, in the Yorkshire Wolds and the Derbyshire Dales, the thistle is almost confined to slopes which face south-south-west. This confirms that in this region, the climate is critical to the species. The south-facing slopes will be warmer than the north-facing ones.

In July and August, some of the rosettes flower. Their female parts are pollinated, fertilisation takes place and the fruits expand and ripen. When the fruits have dried out, they are dispersed.

The microclimate seems to affect the fruit output of the thistle. *The warmer it is, the more viable fruits are produced.* There are three main pieces of evidence in favour of this idea.

(a) Pigott noticed that the thistles produced twenty-five viable fruits per capitulum in south-east Britain but that the fruit production decreased markedly to three to four per head at its northern limit.

(b) The pollinated capitula of plants growing at their northern limit were enclosed in bags of translucent paper, which increased the daytime temperature around the heads by 4–5°C. This considerably increased the numbers of fruits with fully-grown embryos, in some cases ten-fold, when compared with non-enclosed capitula which ripened simultaneously on the same plant.

(c) In another experiment, the flowers were sprayed with water every morning during the period when their fruits were ripening. Their production of ripe fruit was considerably reduced compared with unsprayed control plants growing nearby. The solar radiation falling on the capitula only began to warm up the fruits after it had been used to warm up and evaporate the 0.4–0.6 g of water which accumulated around the florets in each cup-shaped capitulum.

The ripe fruits are dispersed three or four weeks after the first florets open. The weather during this short flowering period is critical. At the northern boundary of the species, flowering usually coincides with long periods of cloudy weather and only a small proportion of the fruits ripen. Three other factors tend to reduce its viable fruit production at the northern limit. Firstly, the thistle produces its flowers about a month later in Yorkshire and Derbyshire than in southern Britain. This tends to reduce the average temperatures which the fruits experience during their development. Secondly, the rainfall is higher towards the northern edge of its range. In wet weather the capitula often fail to open or decay on the ground. Thirdly, in moist weather the heads may be infected by the fungus *Botrytis cinerea*, which makes fruits inviable, destroys viable fruits, or binds them to the receptacle so that they cannot be dispersed.

Why is the stemless thistle so uncommon on north-facing slopes in this area? Even if the occasional viable fruit was carried that far and encountered soil of the right pH, the seedling would probably be outcompeted by the tall grasses which occur on north-facing slopes. If it did survive, it might reproduce vegetatively but would not produce viable fruits. Its demise would be inevitable.

In this chapter we have dealt with the physical, chemical, biotic and historical factors which affect the distributions of species. In the next chapter we shall discuss the factors which determine how many individuals of a species there are.

Population dynamics

Within its habitat range, what determines how many individuals of a species there are? What sorts of fluctuations occur in its population size, and why? Why are some species common and other species rare? Some of the answers to these questions are provided by the study of population dynamics.

If we can understand the factors which influence the population sizes of animal and plant species, we might be able to maintain fish and whale stocks more effectively, reduce outbreaks of pests, and preserve rare species from extinction.

The chapter begins with a description of the changes in the population sizes of species from year to year. Then comes a discussion of the main factors which influence the numbers of individuals in a population, including food supply, territories, predators and parasites. A detailed account of population regulation in the great tit (*Parus major*) shows how a population can be regulated by several different factors. The special features of population regulation in herbaceous plants are dealt with. Finally, two applications of the principles of population regulation are discussed, namely the biological control of pests and the harvesting of an animal population to provide the maximum yield over a long period of time.

THE POPULATION DENSITIES OF SPECIES IN NATURE

When we discuss population dynamics we consider the number of individuals per unit area or unit volume of habitat, in other words, the population density.

In view of the enormous reproductive potential of many species, it is surprising that their population densities are not higher. A female cod, for instance, may produce over a million eggs at once, and a single plant of the common annual weed fat hen (*Chenopodium album*) can make at least 400 000 seeds a year. If all its offspring and descendants survived, a greenfly could produce 600 000 000 000 greenfly in a season, with a total mass of 600 Mg, equivalent to 10 000 men. Charles Darwin recognised this. 'There is no exception', he wrote, 'to the rule that every organic being naturally increases at so high a rate, that if not destroyed, the earth would soon be covered by the progeny of a single pair'. Most zygotes must die before they can reproduce.

Species invading new areas

Occasionally a population expands at the maximum possible rate. This occurs for a short time when a species invades a new habitat, provided that the habitat is suitable and there are no competitors, parasites or predators. Bacteria on an agar plate and yeast cells in broth (figure 30) show a characteristic S-shaped (sigmoid, logistic) curve of population increase with time, under ideal conditions.

The shape of the graphs can be explained in this way. At first, numbers rise slowly, because there are few reproducing individuals in the population and, in many species, some individuals cannot find a mate when the population density is low. There is an exponential increase in numbers. In other words, the rate of increase increases with time. Since the next part of the curve is linear when plotted on a logarithmic scale it is called the 'log phase' in the growth of the population. During this phase the birth rate is much greater than the death rate. Eventually the numbers increase more slowly, and the population density levels out at a maximum value, called the carrying capacity of the habitat for the species, around which it fluctuates between relatively narrow limits.

Why does population growth slow down and stop in this way? At high densities there may be: (a) increased competition between the individuals, (b) an increase in the proportion of individuals which are killed by parasites or predators, (c) in territorial species, the occupation of all the territories in the habitat, or (d) the accumulation of toxic compounds. These factors act

FIGURE 30

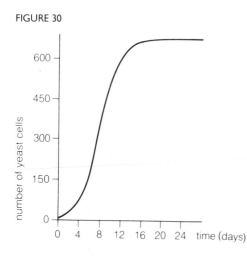

The growth of a population of yeast (*Saccharomyces cerevisiae*) in nutrient broth. Note that the vertical axis is linear. The rate of increase is plotted on a vertical logarithmic scale in figure 31. (G. E. Hutchinson, 1978.)

individually or together to increase the 'environmental resistance'. They lower the birth rate, or increase the death rate, or both, and prevent the population density from rising. This pattern of population growth followed by stabilisation, has occurred frequently in nature. For instance, the population sizes of the European starling in the United States, and of collared doves (figure 31), policeman's helmet (*Impatiens gladulifera*) and rhododendrons (*Rhododendron ponticum*) in the British Isles, have levelled off after expanding rapidly.

Constant population sizes

Once they have reached their carrying capacities, the population densities of many species in nature remain relatively constant. In the long-term of course, over thousands and millions of years, there must have been many climatic, vegetation and evolutionary changes which caused local population crashes and extinctions. Yet in the absence of human intervention such as habitat destruction and pesticides, over periods of fifty to a hundred years, the population densities of most species probably vary between narrow limits.

Many of the censuses available are for species which are economically important pests of human crops. Some species, however, have been counted in apparently 'natural' surroundings. For example, herons in the Thames valley (figure 32) and cowslips in a Swedish pasture (figure 33) both have fairly constant population sizes, which might be maintained for periods longer than a human life span.

In a population of constant size one would expect the birth rate to equal the death rate, or rather, the sum of births and immigrants to equal the sum of deaths and emigrants over a given time span.

In contrast, the population densities of a few species go up and down rhythmically. These cycles have attracted attention.

Cycles

Remarkably regular cycles of abundance and rarity occur in the arctic among the small rodent called the lemming and its predator the arctic fox (a four year cycle), and in the snowshoe hare and its predator the lynx (a ten year cycle) (figures 34–6). In the lemming, the population size at the peaks may be sixteen times that in the troughs.

Several explanations have been suggested for the lemming cycles. One possibility is that over the three years of lemming population increase the individuals are overcrowded and subjected to increasing stress. When the stress exceeds a critical level the lemmings tend to migrate in random directions. This accounts for the stories about lemmings 'committing suicide' by jumping off cliffs or falling into rivers in large numbers. Mass emigration

FIGURE 31
Population size of the collared turtle dove (*Streptopelia decaocto*) in Great Britain from its invasion in 1955 to 1988. The curve can be regarded as sigmoid. (Data from various sources, including G. E. Hutchinson (1978), and data provided by the annual censuses of breeding birds by the British Trust for Ornithology, Tring).

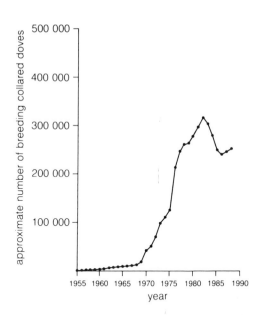

FIGURE 32
The number of breeding pairs of grey herons *Ardea cinerea* in England and Wales 1928–77. (British Trust for Ornithology; C. M. Reynolds, *Bird Study*, **26**, 7–12 (1979)).

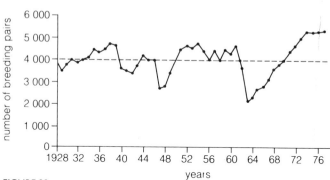

FIGURE 33

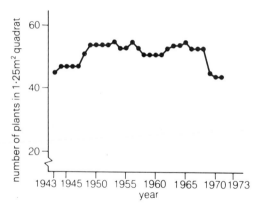

The population size of cowslips (*Primula veris*) in a moist Swedish meadow from 1943–73. (Adapted from C. O. Tamm (1972). Survival and flowering of some perennial herbs, III. The behaviour of *Primula veris* on permanent plots, *Oikos* **23**, 159–166.)

FIGURE 34

lemmings
per
hectare

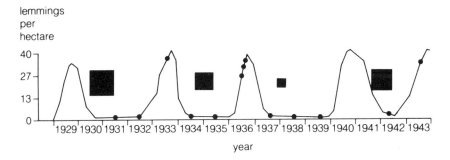

Population density of the lemming *Dicrostonyx groenlandicus* at Churchill, Manitoba. The squares show the invasions of the snowy owl (*Nyctea scandiaca*). (V. E. Shelford (1945). The relation of snowy owl migration to the abundance of the collared lemming. *Auk* **62**, 592–6.)

FIGURE 35
The brown lemming (*Lemmus trimucronatus*). Like many small rodents, this species undergoes regular fluctuations in population density.

FIGURE 36

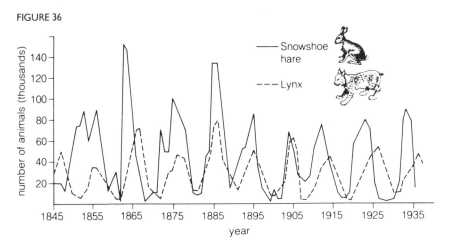

Fluctuations in the numbers of the snowshoe hare (*Lepus americanus*) and the lynx (*Lynx canadensis*) in the Canadian arctic. The numbers show the pelts received by the Hudson Bay Company. (E. P. Odum, 1971.)

lowers the population size drastically, but numbers begin to rise again because there are few lemmings and plenty of food. Another idea involves a four year cycle of plant growth. As the lemmings increase they overeat the grass in their habitat and many of the nutrient ions from the grass become locked up in lemming droppings. In the cold arctic conditions, lemming faeces decompose slowly and the regeneration of the vegetation is limited by the shortage of nutrient ions in the soil. Food shortage stimulates the lemmings to migrate, and the abrupt reduction in lemming numbers allows the vegetation to recover. There is then enough food for the rodents to increase again.

Food shortage stimulates many lemmings to forage for food during the day, thus attracting predators. Lemmings also suffer from more tuberculosis and other bacterial diseases when their numbers are high. One theory suggests that the range of alleles possessed by the lemmings changes throughout the three and a half year cycle, so that genotypes which have the urge to migrate predominate at the population peaks.

The predators on lemmings and snowshoe hares, the snowy owl, arctic fox and lynx, exhibit regular cycles in numbers because of the oscillations in numbers of their prey. Similar, but shorter, cycles occur in laboratory experiments in which a parasite and its host, or a predator and its prey are kept together for several generations. This can be shown by using the orange-eating six-spotted mite, *Eotetranychus*, living on 120 oranges, and a mite which eats it, *Typhlodromus* (figure 37).

The fluctuations in the numbers of the mites, slightly out of phase with one another, can be explained as follows. When the orange-eating mite increases, the predatory mite has more food and breeds more rapidly. Eventually the predator eats such a high proportion of the prey that the prey population declines. Since this reduces the predator's food supply, the number of predators falls. Eventually, such a small proportion of the prey are eaten that their numbers increase again, and the cycle repeats itself. This interaction between the mites has been exploited in the 'biological control' of orange-eating six-spotted mites.

In nature a predator or parasite only undergoes regular cycles of abundance and rarity when it specialises on a single host or prey species. Notice that the artic fox specialises in eating lemmings and the lynx concentrates on snowshoe hares. Those predators or parasites which are generalists (p. 20) can merely 'switch' to alternative sources of food when one host or prey species becomes scarce. Their population densities will remain more constant than those of specialists.

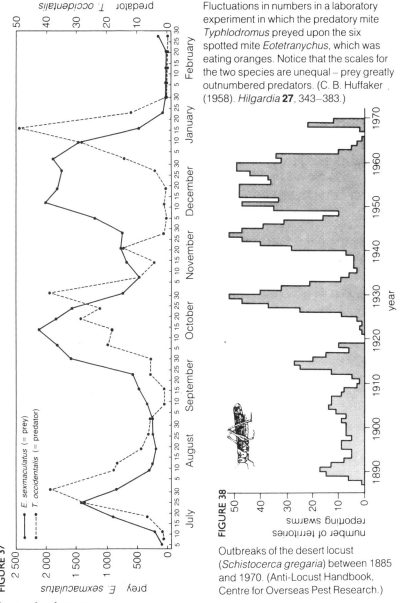

FIGURE 37

Fluctuations in numbers in a laboratory experiment in which the predatory mite *Typhlodromus* preyed upon the six spotted mite *Eotetranychus*, which was eating oranges. Notice that the scales for the two species are unequal – prey greatly outnumbered predators. (C. B. Huffaker (1958). *Hilgardia* **27**, 343–383.)

FIGURE 38

Outbreaks of the desert locust (*Schistocerca gregaria*) between 1885 and 1970. (Anti-Locust Handbook, Centre for Overseas Pest Research.)

Irregular bursts

Some species, most of them insects, undergo irregular population explosions. The abrupt changes in numbers of the desert locust (figure 38), for example, seem to depend on differences between years in weather or climate.

In the years of locust plagues the swarms seem to emanate from areas of high rainfall. Rain provides the moist conditions necessary for the incubation of locust eggs in sand. Wet weather also increases the growth of the plants on which the locust nymphs and adults feed. If insects like locusts experience a population explosion there must be superabundant food and few parasites or predators. If the food supply was limited, competition for food between the individuals would lower the birth rate and raise the death rate. If parasites or predators were present, they would congregate and feast in the areas where locusts were most abundant. In either case, the plague of locusts would not occur.

We ought therefore to distinguish clearly between factors like rainfall which can cause abrupt fluctuations in population size and factors like food shortage, parasitism or predation which tend to prevent the numbers of a species from rising.

DENSITY-INDEPENDENT AND DENSITY-DEPENDENT FACTORS

Locusts exhibit fluctuations in population size which are mainly the result of change in **density-independent factors**. A dry year or a heavy thunderstorm would kill more individuals in a large population of locusts than in a small population. Yet whatever the population size, the same proportion of the individuals would be exposed to the hazard and would be killed by it. The proportion of individuals which dies is therefore, in this case, independent of the population density.

Populations in which most individuals die as a result of density-independent factors tend to oscillate rather wildly in size. The fluctuations appear to be random. In most species these potential oscillations are 'damped' or reduced by density-dependent factors.

Density-dependent factors kill a higher proportion of the population when the density is high than when it is low. When the population density is high, each individual stands less chance of survival because of intraspecific competition for food, competition for territories, or increased predation or parasitism. These factors tend to be density-dependent. The numbers decline and the population size returns to the norm. When there are fewer individuals in the area, there is more food per individual and plenty of territory space. There is also less chance of attack by predators, because the predators are concentrating upon other prey species. At these low population densities each individual stands more chance of survival and, on average, leaves more offspring. This increases the population size and returns it to the norm.

Populations in which density-dependent factors operate like this to stabilise the population size are said to be regulated. Regulating factors tend to increase the population size when it falls below the norm and decrease the population size when it exceeds the norm.

REGULATING FACTORS

In detailed studies of animal populations the main regulating factors to emerge have been competition for food, the formation of territories, predation and parasitism. These will be considered in turn.

Competition for food

Competition for food, particularly intraspecific competition, regulates the numbers of many carnivores and herbivores. In many species, of course, territories, predators and parasites all tend to keep the population size well below the level which the available food supply could support. In the absence of these checks, animals may overeat their food supply, and starve.

This is illustrated by the case of the moose population on Isle Royale in Lake Superior. Some moose (*Alces alces*) swam to the island in 1912 or 1913 and established a population which increased to 1000 in 1921 and 2500 in 1928. At that time the moose had no predators. Naturalists who visited the island in 1934 found many dead and emaciated moose and no plants suitable for them to browse in winter. In their efforts to find food, the moose had even eaten water weeds and nibbled poisonous yew trees (*Taxus canadensis*). The population crashed, reaching 171 in 1943, but then began to increase again. This suggests that in the absence of predators, the moose population was regulated by density-dependent intraspecific competition for food.

In 1948 a pair of wolves (*Canis lupus lycaon*) crossed the ice to the island. They established one large pack and two small ones, feeding almost entirely on moose. Since 1960, the Island has supported constant populations of about 600 moose and twenty-four wolves. The moose population is now regulated by predation, at a level below that which the food supply might support. The size of the wolf population, on the other hand, depends on the abundance of moose. The wolf population is regulated by territories and competition for *its* food supply.

Territories

A territory is an area of suitable habitat which is selected by a male, and defended against other males of the same species. A male with a territory attracts one or more females and breeds with them. Males which cannot establish territories are driven away, and cannot breed.

Most birds and mammals are territorial. Male foxes (*Vulpes vulpes*) for example, use a special gland beneath the tail to scent mark prominent objects around the margins of their areas. Some fish and insects, such as the three-spined stickleback (*Gasterosteus aculeatus*), dragonflies, damsel flies and small tortoiseshell butterflies (*Aglais urticae*), also defend territories. Territories regulate the population sizes of many species, especially those at the tops of food chains.

FIGURE 39

A tawny owl (*Strix aluco*) arrives at its nest carrying a short-tailed vole (*Microtus agrestis*). The main diet of tawny owls is wood mice and bank voles (*Clethrionomys glariolus*). In years in which these prey are scarce, the owls do not breed. Nevertheless, the population density of tawny owls is regulated mainly by territorial behaviour. (Courtesy of G. T. D. Hirons and K. Marsland, A.E.R.G., Zoology Department, Oxford University.)

The tawny owl (*Strix aluco*) is one well-researched species (figure 39) in which territory is an important regulating factor. It lives in mature woodland and eats a diet of small rodents, particularly wood mice and bank voles.

Mick Southern mapped tawny owl territories in Wytham Wood, near Oxford, over twenty-five years. He visited the wood at night. Whenever two owls were heard hooting at one another it was assumed that the dispute occurred at the boundary between two territories. By plotting all these points on the same map a plan of the territories was gradually built up.

The numbers of pairs of owls in the wood increased from thirteen in 1947 to thirty-one in 1955. Then, for the next fifteen years, the population size remained the same. Tawny owls pair for life. Whenever an owl died it was replaced by one of the new birds. Those youngsters which could not establish territories disappeared and presumably found sites outside the wood. Breeding success fluctuated considerably — in favourable years each pair reared two or three offspring, but when bank voles and wood mice were scarce no young were produced. Despite this, the population size of adult owls remained remarkably constant.

Predation

There is only one way to establish whether or not a natural population of an animal or plant species is regulated by a predator or parasite. This is to completely add or subtract the predator or parasite and to record the effects on the size of the population.

We have already seen that the moose on Isle Royale ceased to fluctuate in numbers once the wolves had invaded the island. In this case the predator eats a single prey species. An elegant experiment by Thomas Paine, however, showed that a single generalist predator can simultaneously regulate the population sizes of several prey.

The starfish *Pisaster* is the dominant predator in a community of animals and algae on a rocky intertidal shore in Washington State (figure 40). It feeds on eight species of arthropods and molluscs. In addition there are four species of seaweed, two species of sea anemone, and a sponge with its specialist predator. Paine cleared the *Pisaster* from a zone 8 m across and 2 m down the shore, and left a similar area untouched as a 'control'. In the control, a year later, all fifteen species and the starfish remained. In the zone without *Pisaster*, the barnacles began to spread. Three months after *Pisaster*

FIGURE 40

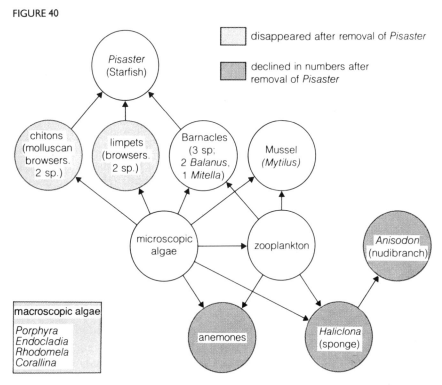

Food web for the starfish *Pisaster* on a rocky intertidal shore in Mukkaw Bay, Washington State, showing the species which were eliminated when *Pisaster* was removed from the community. The disappearance of *Pisaster* was followed by a population explosion of barnacles and mussels. (R. T. Paine (1966). Food web complexity and species diversity. *American Naturalist*, **100**, 65–75.)

were removed, the barnacle *Balanus glandula* covered 60–80% of the surface. The chitons, limpets, and seaweeds disappeared, and the anemones and the sponge became scarce. Presumably they were all outcompeted by barnacles. The removal of the predator thus caused a population explosion amongst some of its prey, a reduction in the number of species in the community from fifteen to eight, and a considerable change in the relative numbers of the remaining species.

Under natural conditions the starfish probably keeps down the numbers of all the chitons, barnacles and limpets on which it preys. If a prey species were to become particularly abundant, *Pisaster* would eat it preferentially and eventually make it scarce. It would then begin to eat the next most abundant prey species. *Pisaster* maintains such low population densities in its prey that there is enough space in the rocky intertidal community for seaweeds, anemones and sponges. This example illustrates the paradox that a prey species like a chiton or a limpet may remain in a community if its predator is present, but may be eliminated if its predator is absent.

Many other generalist predators, in a wide range of communities, may act like *Pisaster* in regulating the population sizes of several prey species at once. Experimental evidence is lacking. There is some good circumstantial evidence, though, that predators and parasites can have massive effects on the numbers of their prey or host species. In 'biological control' (p. 70) a predator or parasite is introduced by humans into a population of a pest species, in an attempt to reduce its density to an acceptable level. In most cases, for instance when moths were introduced into Australia to reduce the prickly pear cactus (figure 27 page 42), or ladybirds (*Rodiola cardinalis*) were used to control a scale insect (*Icherya purchasi*) on citrus trees in California (p. 71), the pest populations were reduced to a thousandth or a hundred thousandth of their previous levels. It therefore seems likely that in their native habitats, pests such as the prickly pear cactus and scale insects are 'regulated' by predation and that is why they are not particularly abundant.

Specialist herbivores in the tropics may be responsible for the low densities of individual species of rain forest tree. One of the great mysteries of biology is the large numbers of plant species in tropical rain forests. In a study of a tropical forest in Malaysia, for instance, 375 different tree species were recorded in 23 ha. Each species was represented by only 7.4 individuals, on average. Yet in the same area of temperate forest in the British Isles there might be eight or nine tree species.

Dan Janzen investigated over a hundred species of bruchid beetles which feed on the seed pods of tropical rain forest trees in Costa Rica. Each beetle specialised on a different tree species. It could not attack the other trees because they all had chemical defences. In other words, most trees were resistant to all the beetle species except one. The bruchids attacked the seeds on the ground, so that virtually no seeds survived when they came to rest near the parent tree. This ensured that individual trees of the same species are widely separated, since seeds were only likely to germinate, and seedlings to survive, some way away from their parents. If this situation is widespread in the tropics it may account for the coexistence of numerous species of rain forest trees at low densities.

Parasitism
The parasites which are most important in regulating animal populations are firstly, the parasitic insects, and secondly, disease-causing organisms like fungi, bacteria and viruses. These will be considered in turn.

Over 100 000 species of insects, making up nearly a tenth of the animal kingdom, make a living by parasitism. These are mainly small wasps (*Hymenoptera*) and flies (*Diptera*), sometimes hardly visible to the naked eye. The majority attack other insects, and are fairly host-specific. Many attack plants and form galls, which are chambers, formed by a disruption of normal plant growth, in which the young stages of the insect can develop in a protected environment with an abundant food supply. Other parasitic insects favour spiders, centipedes, millipedes, molluscs, annelid worms or vertebrates.

Many parasitic insects resemble predators more than parasites. The females, after mating, lay their eggs inside or on their hosts. The larvae which hatch out usually eat the internal organs of the host and kill it, before pupating and hatching out as adults. These species differ from other parasites in that they kill and eat their hosts.

There is a great deal of evidence that parasitic insects can regulate the sizes of their host populations. Laboratory experiments, stimulated by the importance of parasites as agents of biologial control, show that when the hosts are thick on the ground the ability of the parasite to destroy them is limited by both the restricted egg-laying capacity of each female parasite and the time taken to lay eggs in an individual host. Nevertheless, the parasites flock to the areas of the highest host density and this enables them to regulate their host populations in a density-dependent manner.

Parasitism by fungi, bacteria or viruses may sometimes regulate the population sizes of their hosts. The rate at which a parasite spreads through a population depends on its generation time, reproductive output, dispersal mechanisms, the average density of the host, and the genetic diversity between the individuals in the host species, some of which may be resistant to attack. In a sparse host population, the chance that a parasite will produce offspring which will find a new host may be so small that a localised outbreak of the parasite will soon disappear. On the other hand, when there are more potential hosts per unit area, the parasite may spread from one to another so rapidly that there is positive feedback. The number of infected individuals increases exponentially with time. Eventually this will produce an epidemic.

Such epidemics are rare in nature. It seems that most wild populations have evolved parasite tolerance or resistance. The genetic differences between the individuals in a sexually-reproducing host population cushion the population against parasite attack. Only a few host individuals are susceptible and this lowers the effective density of the hosts.

Most epidemics have occurred in dense monocultures of plant species. In many cases the genetic diversity of the host species is limited, and the disease has been newly introduced to the area so that the host has not encountered it before. In 1971, for example, an aggressive strain of *Ceratocystis ulmi*, the fungus which causes Dutch elm disease, was introduced into Britain on logs of the rock elm imported from Canada. It spread rapidly from Harwich, London, Bristol and Southampton, its spores being carried by the elm bark beetles, *Scolytus* spp. Of the twenty-three million elms (*Ulmus* spp.) south of a line from Bristol to the Wash, some fourteen million had been killed by 1978. There is limited genetic diversity in the populations of most hedgerow elms in Britain. Elms were established in hedgerows by planting the suckers from elm trees growing nearby.

It is often difficult to show that parasitism acts as a regulating factor. The red grouse (*Lagopus lagopus scoticus*), for example, lives on heather moors on the acidic uplands of the British Isles and is shot for sport. Grouse shooting is big business. Millions of pounds have been spent on research. At first attention focussed on nutrition and territory size. To maintain the nutritious young heather shoots on which grouse feed, the moorland is burnt on a rotational system, each strip burnt every eight to fifteen years. Despite the improvements in heather management, however, grouse populations have declined in Scotland and are cycling in the north of England. Attention has now shifted to the role of a parasite which kills grouse, a nematode worm *Trichostrongylus tenuis*.

Several examples have been mentioned in this section of populations which are regulated by food supply, territoriality, predators or parasites. Some species, however, are regulated by several of these factors. Populations of the great tit have been studied in detail at Wytham Woods near Oxford for over thirty years, and illustrate this point well.

POPULATION DYNAMICS OF THE GREAT TIT

During the last thirty years the population of great tits in Marley Wood has remained fairly stable (figure 41). The graph obscures the fact that there is always a marked difference in numbers between the summer and the winter. The young birds which leave the nest in May have a short expectation of life.

Over the years, hundreds of nest boxes have been attached to trees in the wood and the great tits prefer to nest in them. This makes it easier to observe and count breeding pairs, and to follow the numbers of eggs, chicks and fledged young. The young birds are ringed. At the time of writing (1989) every pair of great tits in the 480 ha wood nests in a nest box and every great tit in the wood has been ringed.

In some studies the young birds have been weighed at intervals, so that the relationship between birth mass, fledgling mass and the probability of survival to a particular age can be determined. Several of the nest boxes have flash cameras built in to the back so that whenever a bird enters the nest hole it is automatically photographed. The photographs record the time at which the bird entered the hole. They also allow any food items in its beak to be identified (figure 42).

FIGURE 41

The numbers of pairs of great tits breeding in Marley Wood, Wytham Woods, Oxford, 1947–1985. (C. M. Perrins (1980). The great tit, *Parus major. Biologist* **27** (2), 73. Subsequent data by courtesy of Dr Perrins.)

FIGURE 42

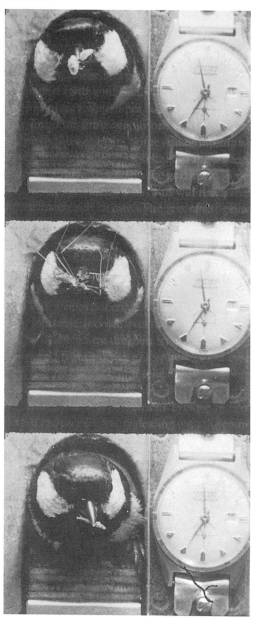

Three frames from a film taken with an automatically triggered flash camera inserted in the back of a great tit nest box. Great tits mainly eat insects in summer. Note how rapidly the visits were made.

Right Automatic photograph of a weasel backing out of a nest box with a nestling of a great tit. Predation by weasels is one of the density-dependent factors which regulates the population density of great tits in Wytham Woods, near Oxford. (Both courtesy of C. M. Perrins, from *British Tits*.)

The production of young

Male and female birds pair up in autumn and the males begin to establish territories in the following January. They deter intruders by singing and threat displays. The males and females that fail to gain a territory leave the wood.

How can one prove that the males compete for territories and the losers are banished? Krebs carried out an ingenious experiment. He shot and killed some of the pairs. He worked in an isolated wood in which fourteen to sixteen pairs of great tits bred every year and most of the birds were ringed. In 1968, seven pairs were shot in mid-March when territories had been established, and five new pairs settled within six days. In 1969, six pairs were killed and four new pairs replaced them within three days. Where did the replacements come from?

In 1969, most of the great tits within 800 m of the wood had been colour-ringed. The new settlers had come from nearby hedges where they had already formed territories. Hedges seem to be a second rate habitat for great tits. Only 20% of their nests in hedges produced young birds, perhaps because the nests were exposed to predation. It is not surprising that the males in hedges move into the wood at the earliest opportunity.

The birds originally nested in holes in trees but now nest in nest boxes. The females lay a 'clutch' of about ten eggs in late April. The number of eggs laid varies considerably. More eggs are laid per nest when the birds are sparse than when they are abundant. The number of eggs per clutch seems to be a density-dependent factor which regulates the size of the population.

One egg is laid each day. They all hatch out over a period of two or three days nearly two weeks after the last egg has been laid. When food is abundant, that is, when the parents make frequent visits to the nest during the day with caterpillars in their beaks, most of the chicks increase rapidly in weight. When food is scarce, however, only the first chicks to be born put on weight. The others die without leaving the nest. In this way the population density of the tits is adjusted to the food supply.

Another factor which acts at the nesting stage and affects the population size is predation by weasels (*Mustela nivalis*). The weasels smash the eggs and kill the chicks. John Krebs found that when nests were <45 m apart, 23% were attacked by weasels. When the nests were >45 m apart, 11% were attacked by weasels (figure 43). Weasel predation is therefore density-dependent. It kills a higher proportion of the nestlings when the nests are close together, the territories are small and great tits are abundant.

The fate of the young which leave the nest

The chicks take two weeks to grow feathers. They then begin to make practice flights. They are fed by the parents for a further two weeks. At the

FIGURE 43

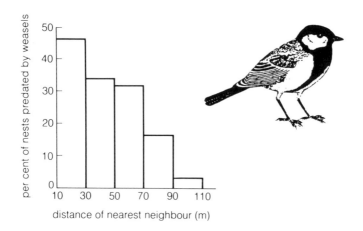

distance of nearest neighbour (m)

The influence of territory size of the great tit on the risk of nest predation by weasels. (After J. R. Krebs (1971). Territory and breeding density in the great tit, *Parus major* L. *Ecology* **52**, 2.)

beginning of this period caterpillars of species such as the winter moth and green tortrix (*Tortrix viridana*) are abundant. In mid-May or early June most of the moth larvae pupate. This greatly reduces the food supply.

This period just after the young have left the nest seems to be critical for the survival of the brood. Not only may food be scarce but also there may be serious predation from sparrowhawks in some years. Sparrowhawks were absent from Wytham Woods from 1959 to 1968 but there are now nine pairs nesting there. Very few great tit broods survive within 60 m of a sparrowhawks' nest, particularly when voles and mice, the sparrowhawks' alternative prey, are scarce. In early summer hundreds of plastic rings from the legs of tits can be found in sparrowhawk droppings.

From July to September there is competition between the tits for greenfly in the tree canopy. The adults which have just bred survive this competition rather better than their offspring. The ratio of adults to children is about one to three when the fledglings begin to fend for themselves but about one to one by the end of summer. The youngsters which die tend to be those which weighed the least when they left the nest (figure 44). Competition for food in summer is yet another density-dependent factor which tends to regulate the population size. On average, from a clutch of ten eggs, only three offspring survive to the end of September.

FIGURE 44

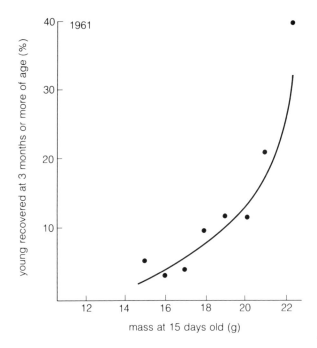

Long-term survival of young great tits in relation to their nestling masses on the 15th day after hatching. Survival was measured by the percentage of young, of any given mass, which were later recovered alive. (C. M. Perrins, 1980.)

Competition for food is relaxed in autumn, when the tits descend to the forest floor. They feed on beech and the insects associated with fallen leaves.

Nevertheless 70–80% of the individuals may die in a hard winter. Presumably the deaths would be density-dependent if they died of starvation as a result of competition for food, and density-independent if they froze to death.

It seems therefore that no single factor is of overriding importance in regulating the population size of great tits in Wytham Wood (figure 45). A large population encounters, at any time of year, some factor which tends to reduce its size. Small populations meet less resistance, and expand.

FIGURE 45

Factors affecting the population size of great tits in Wytham Woods, near Oxford.

(a) Number of eggs laid by female (clutch size). (*d–d*).

(b) Proportion of eggs which hatch, which mainly depends on the intensity of predation by weasels. (*d-d*).

(c) Competition amongst nestlings for the food supply gathered by their parents? (*d–d?*).

(d) Predation of nestlings by weasels. (*d–d*).

(e) Body mass of fledglings when they leave the nest influences subsequent survival. (*delayed d–d*).

(f) Predation of fledglings and older birds by sparrowhawks (*Accipiter nisus*). (*delayed d–d?*).

(g) Competition for scarce aphids and small insects in the tree canopy in summer? (*weak d–d?*)

(h) Food availability (e.g. beech mast) during winter. (*d–d*).

(i) Winter weather. (*d–i*).

(j) Territory formation. (*weak d–d*).

(k) Competition between blue tits (*Parus caeruleus*) and great tits (*probably d–d*).

Key

d–d = density-dependent regulating factor.

d–i = density-independent.

POPULATION REGULATION IN FLOWERING PLANTS

There are three main differences between the life cycle of a flowering plant and an animal. Firstly, in many plant species most of the individuals are invisible, since they are embedded in seeds underground. Secondly, the seeds can survive underground for tens or hundreds of years before they germinate. Animals rarely undergo such lengthy periods of dormancy. Thirdly, many plant species reproduce asexually, by vegetative reproduction, as well as sexually. This often makes it difficult to define and count individual plants.

Partly because of these difficulties, less is known about population regulation in plants than in animals. The fluctuations in numbers of plants have been studied mainly in plants growing in grasslands, on sand dunes, in vegetable gardens and in crops. Most investigations have concentrated upon short-lived plants, rather than trees. Thus, after a discussion of the stages in the life cycle at which a plant population can be regulated, creeping buttercup, a short-lived perennial, will be dealt with in more detail.

The life cycle of a flowering plant

The population size may be regulated by density-dependent factors at several different stages. Let us begin with the mature plant.

At the mature plant stage the production of seeds or vegetative offspring by individual plants may be reduced at high densities. A single plant of the garden weed groundsel (*Senecio vulgaris*), for instance, might be 3 cm high with a single capitulum when grown under competitive stress but 40 cm high with tens of capitula when grown on its own in sunny, moist nutrient-rich conditions. This is an example of 'phenotypic plasticity', the ability of an organism under environmental stress to adopt a growth form which adapts it to the conditions in which it is growing. In studies of highly plastic annuals and ephemerals it is therefore useless to count individual plants without recording their size as well.

Mature plants may also attract more parasites and herbivores at high than at low densities. In plant communities which are grazed by rabbits, sheep or cattle, the increase in one species in the community may be counteracted by a tendency for this species to be grazed more frequently, particularly if it is palatable. This prevents a single species from becoming dominant in the grassland and maintains a wide variety of plants. In fact the generalist herbivore in the grassland is behaving like *Pisaster* in its rocky inter-tidal community (p.57).

The seeds produced may be eaten by predators or infected with parasites. A high proportion may fall on 'stony ground' instead of the places, called microsites, where the environmental conditions are most favourable for seed germination and seedling establishment. If the microsites available for (a) seed germination, and (b) the establishment of vegetative propagules, are limited in number, the population will be regulated in a density-dependent manner.

In a vegetable garden or an arable field, there may be 40 000 buried viable weed seeds per square metre. The number of dormant seeds in a soil sample can be estimated by churning up the sample at intervals, keeping it moist in a shallow tray and identifying the seedlings which emerge. An alternative method is to separate the seeds by flotation and to estimate their viability by a chemical colour test. If the plants above the ground are removed or die, they are rapidly replaced from this seed bank. The presence of the seeds in the soil thus provides a buffer which prevents the species from being eliminated from the habitat.

Once the seeds have germinated, there may be considerable seedling mortality. In dense populations of seedlings, intraspecific competition causes many seedlings to die. Dense populations are also more likely to be infected by fungi or attacked by predators.

Mature plants, especially of the same species, may begin to compete with one another. The result can either be density-dependent mortality (p.54), or else a reduction in growth rate. This reduction in growth rate takes the form of a reduced seed production, or a decrease in the number of vegetative offspring produced.

Some of these principles are illustrated by the results of studies on the population dynamics of the creeping buttercup.

THE CREEPING BUTTERCUP

The creeping buttercup (*Ranunculus repens*) is a perennial with vigorous vegetative reproduction. Its population dynamics were studied by José

Sarukhan in a grazed coastal meadow near Bangor, Wales. The creeping buttercup flowers in June but each flower produces on average only two or three viable fruits. Of these, over half are eaten by predators on some plots, and only 10% germinate. The life expectancy of a plant arising from seed is only 0.2–0.6 years. Vegetative reproduction is much more successful. A single rosette can produce up to five stolons (overground stems), and each stolon bears three or four nodes about 5 cm apart. Each node has one or two leaves, and a pair of adventitious root meristems which grow downwards and anchor the node to the ground. The daughter plants become independent when the stolons wither and break; they are similar in every way to the surviving older plants. Their life expectancy is 1.2–2.1 years.

Every individual of the species were mapped in permanent quadrats at frequent intervals. This is a very lengthy and time-consuming business. In fact it is rumoured that the driver of the London–Holyhead express, seeing Sarukhan stretched out amongst the buttercups in the same place as he was a few hours before, stopped the train to inspect the corpse!

In one of the plots there were 117 individuals at the beginning of the study and 139 individuals two years later. Yet only thirteen plants had persisted for the whole two years. During that time, 244 new plants had arrived and 22 had died. Most of the recruits had been produced by vegetative reproduction. In the study as a whole, the creeping buttercup plants produced 115 vegetative daughters m^{-2} in one year and 114 m^{-2} in the next.

Indeed, two aspects of vegetative reproduction seem to regulate the population size. Firstly, the probability that a creeping buttercup will produce vegetative offspring is reduced at high densities (figure 46). Secondly, when the life expectancy of these vegetative daughters is plotted against the initial number of plants per quadrat (figure 47) it is clear that the daughters born into sparse populations of the creeping buttercup have a greater chance of survival than daughters born into dense populations. This is a density-dependent effect. Small populations gain more recruits than large ones.

These two density-dependent factors seem to act together to 'regulate' the size of the population. When numbers are small a high proportion of the plants reproduce and their vegetative offspring last a long time. When numbers are large, few plants reproduce and vegetative offspring die more rapidly.

It might seem that studies like this, on wild species of animals or plants, are merely of academic interest. This is not so, for the knowledge won and the principles gained can be applied to problems which are economically important.

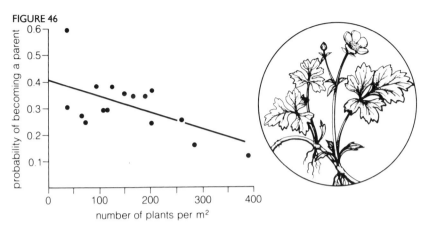

FIGURE 46

Creeping buttercup. The vertical axis represents the chance that a plant present in April will produce at least one offspring by vegetative reproduction. As the density increases the probability that a plant will become a parent declines. (J. Sarukhan (1974). Studies on plant demography: *Ranunculus repens* L., *Ranunculus bulbosus* L. and *R. acris* L. II. Reproductive strategies and population dynamics. *Journal of Ecology*, **62**, 151–77.)

FIGURE 47

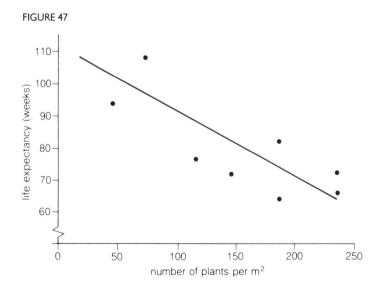

Creeping buttercup. The vertical axis shows the life expectancy of a plant which has been produced by vegetative reproduction. The life expectancy declines with increasing density of creeping buttercups. Plants in sparse populations live longer. (J. Sarukhan and J. L. Harper (1973). *Journal of Ecology*, **61**, 675–716.)

APPLIED PROBLEMS

There are two applied problems in particular which need to be solved on the basis of an understanding of the population dynamics of individual species. These are the 'biological control' of pests and the 'sustained yield' of fish.

Biological control

A species which invades a new area may escape from the predators or parasites which regulated its population size in its native habitat. It may undergo a population explosion, and become a pest. In this case its numbers may sometimes be reduced by introducing one of its predators or parasites (called here 'predasites' for short) into the new area. Control agents which have been used include viruses, bacteria, parasitic insects, and predatory insects.

Biological control is therefore an alternative to the application of pesticides. Both methods have their advantages and disadvantages.

The advantages of biological control over pesticide application are as follows.

(a) Biological control agents tend to be highly specific; they attack a single pest species. Pesticides are non-specific and probably kill many useful species as well as harmful ones. An insecticide sprayed on a crop, for example, may kill ladybirds (*Adalia* spp. and *Coccinella* spp.) as well as the greenfly which they prey on.

(b) The initial research to find a biological control agent takes time, money and manpower. Yet the cost of developing a new pesticide is usually even greater, because of the experiments which are now required to convince governments that the pesticide will have no long-term toxic effects on humans or wildlife. Biological control agents cannot cause pollution.

(c) Once a predasite is released into the wild, and begins to control the pest at an economically tolerable level, it may prevent pest infestations for hundreds of years at no extra cost. Pesticides, many of which are petroleum products, are costly to make and buy, and must be paid for by the farmer year after year.

(d) Imagine that the pest species mutates and becomes either unpalatable or resistant to its predasite. There will then be natural selection in the population of the biological control agent for individuals which can still attack the pest. In fact the only individuals which will have their genes represented in the next generation are these which can still attack the mutated race of the pest species. A biological control agent therefore has a genetic flexibility which enables it to counter any genetic trick perpetrated by the pest. A pesticide has no such flexibility. Its application causes natural selection for pesticide resistance amongst the individuals in the host population. Once resistant strains of the pest have evolved, they can, like DDT-resistant malarial mosquitoes in India, undergo a population explosion, unaffected by the pesticide.

Despite these advantages, biological control has only been used in a few cases, whereas pesticide application is a major activity on which much of our agricultural production depends. Pesticides have three main advantages over biological control.

(a) Pesticides kill large numbers of pest individuals in a short time. Biological control may take months to become effective, by which time a crop might have been destroyed.

(b) Since most herbicides or insecticides attack a broad spectrum of pests, they are effective whichever species of plant or insect happens to undergo a population explosion in that particular year. Most biological control agents are too specific. If a species which was very closely-related to their normal host or prey underwent a population explosion, they would be unable to attack it.

(c) Pesticides usually eliminate the pests for which they are intended. Biological control is often unsuccessful. Of 223 attempts at biological control up to 1969, only forty-two achieved complete success.

Biological control is therefore not a universal panacea for pest outbreaks, but it has proved useful in many important cases and it does not cause environmental pollution. Following a pest outbreak, the natural enemies of the pest are sought in its country of origin. These enemies are imported, studied under strict quarantine, screened for possible harmful effects on beneficial insects or crops, and bred. After a successful introduction of the predator or parasite to the pest population, the erstwhile pest and its enemies may coexist at very reduced densities for many years.

The first pest to be controlled in this way was the cottony cushion scale insect *Icerya purchasi*. When the Californian citrus fruit industry was established in the early 1880s the orange, lemon and lime trees were imported from Australia. The scale insect, which had reached California a few years previously without its predators, invaded the citrus trees and rapidly expanded in numbers. It reduced the productivity of the trees by sucking out the contents of the phloem, and by plastering the leaves with excreta which was rapidly invaded by parasitic fungi. An American biologist was sent to a region of Australia where the scale insect was not a pest, to look for its natural enemies. He returned with a ladybird, the vedalia beetle *Rodiola cardinalis*. In 1889 five hundred vedalia beetles were released on citrus trees which were heavily infested with the scale insect. They rapidly brought the pest under control (figure 48).

The only serious outbreaks of the cottony cushion scale insect since then have been after the citrus trees have been sprayed with insecticide

FIGURE 48

The vedalia beetle (*Rodiola cardinalis*) feeding on cottony cushion scale insects (*Icerya purchasi*) in California. The scale insects look like 'cream cakes' in the photograph. The ladybirds introduced from Australia in 1889 saved the citrus industry by rapidly reducing the numbers of scale insects. (Courtesy of Dr P. De Bach.)

against other pests. Although insecticides kill both the scale insects and the vedalia beetles, the scale insects, in the absence of predation, reproduce rapidly and undergo a population explosion.

A very different method of biological control can be used to eliminate those insect pests in which the females only mate once. In many insects the spermatheca of the female is filled by sperm at the first mating. These sperms are used by the female to fertilise her eggs throughout her egg-laying life, and she does not mate again. In the radiation–sterilisation technique, the natural insect population is swamped by males which have been sterilised by radiation. If a female mates with a sterile male she will produce no offspring. The birth rate declines dramatically, and the population size is reduced year by year.

In the southern United States, for instance, the screw-worm fly (*Callitroga* spp.) is a major pest, particularly since the natural population is augmented by flies which come across the border from Mexico. It lays its eggs in the hides of cattle, and frequently kills them. In a massive 'fly factory', more than 50 Mg a week of blood and meat are used to produce 100 million sterilised fly pupae which are parachuted into the countryside in cardboard containers. This method has virtually eliminated the screw-worm fly from the United States. Of course this very expensive technique can only be used to control a few pests. It only applies to economically important insects in which the females only mate once.

Sustained yield

There are many natural populations of animals and plants which are harvested by humans. If too many individuals in the population are harvested at once, there is the danger that the population will become extinct and the resource will disappear. An understanding of the population dynamics of each species is most valuable in deciding how much predation by humans the species can withstand.

The results of over-harvesting are apparent in the disappearance of many of the largest whale species. Modern purse-seining whaling ships can locate whales easily with echo-sounding equipment. Rapid communications between ships, which hunt in packs, ensure that the human predators congregate in places where their prey is abundant. As a result, most of the larger baleen whales have almost disappeared (figure 49). With the decline of the blue whale, the largest animal on Earth, attention shifted to fin whales. When fin whales became scarce in the 1950s, whalers began to catch sei whales, which are even smaller. Attempts to make and enforce international agreements on the number and species of whales which should be caught have been unsuccessful. Holland and Norway have sold their whaling fleets. The Russians and Japanese have considerably reduced the scale on which they whale.

Several important fisheries have also collapsed and the likely cause is overfishing. The Alaskan salmon fishery, for instance, began in 1880 and catches rose to a peak of 8 500 000 cases of salmon, each weighing 22 kg, in 1936. Since then there has been a continuous decline in the catch despite the fact that more and more boats have been engaged in fishing and the demand for salmon has remained high.

FIGURE 49

▲ fin whale *Balaenoptera physalus*
–○– sei whale *B. borealis*
–●– blue whale *B. musculus*

Total catches of the major species of baleen whales between 1920 and 1970. Whaling activity was considerably reduced during the second world war. (Gulland, 1970.)

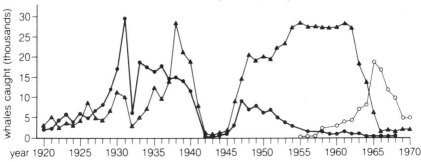

What can be done to prevent disasters like this? Ideally, the way in which a species is harvested should be based upon a thorough understanding of its population dynamics. Such an understanding is difficult to achieve in a vast volume of water like the sea, for species which often migrate over considerable distances. Some of the simplest considerations to be borne in mind in deciding on a harvesting strategy for a fish population are mentioned below.

(a) The best policy, in terms of figure 50, would be to harvest from the usable stock each year the maximum number of fish which could be replaced from growth and recruitment by the following year.

(b) In terms of the typical S-shaped curve of population increase (figure 30, p. 48), the fish population increases most rapidly with time in the middle of the graph, when the population is in the 'log phase' of growth. A large harvest at that population density could result in the rapid replacement of all the harvested fish. On the other hand, if the same number of fish were harvested from a population at high density, they might take years to replace, since near the top of the graph the population increases slowly with time.

Thus to obtain the maximum yield over a long period of time, the population which is harvested must be kept at less than maximum density.

(c) Natural fluctuations in their populations may occasionally reduce fish stocks to dangerously low levels. In 1972 there was a freak movement of warm tropical water into the cool, nutrient-rich water along the coast of

FIGURE 50
The factors affecting the mass of fish available to be fished, the 'usable stock'. (After Ricker.)

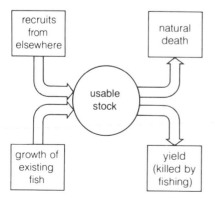

FIGURE 51
The relationship between the development of young and migrations in a population of fish. (Adapted from D. H. Cushing (1968). *Fisheries biology – a study in population dynamics*. University of Wisconsin Press, Madison.)

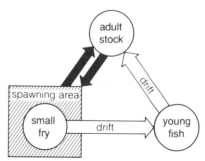

Peru. The anchovy population declined but fishing continued at the usual level. The result was the collapse of the Peruvian anchovy (*Engraulis ringens*), the largest fishery in the world (12.3 Gg in 1970, 18% of the total world fish catch).

(d) The range of sizes of fish which are caught depends on the mesh size of the nets which are used to catch them. A net of very large size will only capture the very largest fish in the population. If most of the fish of breeding age are caught, then in terms of the simple diagram in figure 51, spawning and the production of young fish might be reduced, and the whole population might decline. If the mesh size is too small, so many young individuals will be removed that recruitment to the adult stock slows down. One continual problem is that fish of many species, of varying size, are caught in the same net at the same time. The correct mesh size for one species is not the correct mesh size for another.

(e) Other factors which are important in the harvesting of fish populations are the sites at which fishing occurs, the time of year they are fished, the numbers of boats allowed onto the fishing ground and the rate at which they catch fish.

Most rational people believe that a natural exploited population should be treated like a bank account from which we should take the interest, leaving the capital intact. The interest rate should depend on our knowledge of the particular species, and it should be the maximum yield which is likely to be sustained over a long period of time.

In this discussion each species has been considered in isolation. Yet every species exists within a community of other plants and animals, many of which affect is distribution, behaviour, life-history and abundance. The next two chapters, on succession and ecosystems, respectively, are devoted to the ecology of whole communities.

Biomes, communities and succession

Plants, animals, and micro-organisms live in communities, like the oak forest described in the first chapter. Some types of community extend over very large geographical areas. These are called biomes. The sort of biome which potentially exists at a particular place seems to be determined largely by the conditions of rainfall and temperature which prevail (figure 52).

The climax communities in some biomes, such as tundra, coniferous forest and tropical rain forest, sometimes extend unbroken over thousands of square kilometres. Yet much more of the land surface is covered by plant and animal communities in various stages of succession.

Succession in the predictable, gradual change in the plant and animal communities at a site, eventually leading to the development of a climax (chapter 1). Some successions occur on a small scale. There is a succession of animal species in a cowpat, a succession of vultures which attack a corpse, and a succession of fungi on a rotting log. This chapter, however, concentrates upon large scale successions. Those which occur in newly bared rock, mud or sand are called primary successions. Those which occur where the plant cover has been removed and the soil has already been formed are known as secondary successions.

As an example of succession, consider a newly bared rock surface on a mountainside. It might be invaded first by lichens and by mosses, then by grasses, then by small shrubs of the heather family and finally by taller shrubs and small trees. Eventually a climax woodland would become established. During this process, which might take a thousand years, a soil would gradually be formed. The features which this successional process has in common with all large scale successions are listed below.

(a) The order in which the groups of species achieve dominance, and the species composition of the climax which is ultimately formed, is predictable and repeatable. It depends upon the species composition of the surrounding vegetation, the rock or soil type, and climate.

(b) One set of plant species modifies the soil and microclimate in a way which favours the establishment of the next group of species. The new

FIGURE 52

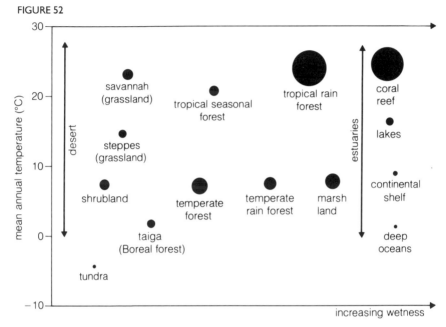

The approximate relationship of biome types to climate. The diameter of each circle is proportional to the net primary production per unit area (see figure 59). (After F. L. Milthorpe and J. Moody (1974). *An Introduction to Crop Physiology.* (Cambridge University Press, with additions from Whittaker and Lieth.)

group of species then achieves dominance and ousts the old one. This process continues until the climax stage is reached.

(c) During succession the height and mass of the vegetation increases. As the vegetation becomes taller, it exerts an increasing influence on the micro-climate within it.

(d) As succession proceeds, the plants and animals modify the soil. With time the soil becomes deeper and its organic matter content increases.

(e) As feeding and structural niches for animals appear and disappear, the fauna gradually changes.

(f) During succession, the number of insect species in the community increases.

Since primary and secondary successions are such different processes, they will be discussed separately.

PRIMARY SUCCESSION

Newly bared surfaces on which succession can occur are continually created by natural processes. These processes include the retreat of glaciers, the erosion of rocks on mountains, the deposition of volcanic lava, and the accumulation of wind-blown sand or water-borne silt. All these surfaces are ripe for colonisation by living organisms and successions inevitably occur on them. As four examples of primary succession, we shall discuss succession on bare rock, in lakes, on sand dunes and in salt marshes.

Succession on rock surfaces

Successions, albeit slow ones, occur wherever bare rock is exposed, for instance on sea cliffs, mountain screes, or rocks bared by glaciers.

The first species to become established on the rock are lichens and mosses, since only they can withstand the desiccation, the low level of nutrient ions and the extremes of temperature which occur on naked rock. Lichens obtain their water and most of their nutrient ions from rainfall. Their rhizines, root-like projections, enter cracks in the rock, and the lichens themselves are pressed closely to the surface. Lichens may begin to break down the rock when compounds like lichen acids, released from the thallus, may combine with ions in the rock to make them soluble, thus helping the rock to break down. Lichens themselves may only play a minor role in rock breakdown, but all the time the newly bared surface will be weathered. The organic matter, deposited by the lichens when they die, will accumulate. This increases the water-holding capacity of the rock surface. It also holds nutrient ions, which could be used by subsequent invaders.

The lichens and mosses tend to be colonised by tufted, upright moss species. The mosses take up water and ions from below and need a water supplying substrate to survive. The lichens, having created conditons more suitable for moss, will soon be shaded out. As the mosses deposit more organic matter, and the weathering of the rock surface begins to create small mineral soil particles, a shallow soil is formed. This is colonised by bacteria, fungi, nematode worms, arthropods and molluscs. These organisms increase the rate at which organic matter is broken down. They form humus and release nutrient ions.

When the soil depth and water-holding capacity have increased, ferns and higher plants may invade the primitive soil. The pressures exerted by their powerful roots may increase the rate at which the lumps of rock weather to form mineral soil particles. Ultimately, as the soil deepens still further, the shrubs and then trees can establish themselves, provided that the climatic conditions are favourable. Once a forest climax has appeared, it may be able to perpetuate itself for thousands of years if the climate remains the same and there is negligible human activity.

This process, which takes at least several hundred years, has probably created the majority of soils over bedrock.

Succession in water

Freshwater lakes may undergo a plant succession which results in a climax forest on dry land! The stages in this slow but dramatic process are depicted in figure 53. Successions like this occur in neutral to alkaline lowland lakes which are rich in nutrient ions, but not in acid mountain tarns which are deficient in nutrients.

Suppose that the lake has an inflow stream, which is likely to carry silt, and an outflow stream. The succession begins when the margin of the lake is colonised by reeds, *Phragmites australis*, or bulrushes, *Typha latifolia*. Both these species spread rapidly by underwater rhizomes and can survive continuous submergence. Silt and clay particles from the inflow stream accumulate around the stem bases and rhizomes of the reeds, and as they do so, the level of silt around the edge of the lake gradually rises.

Conditions soon become suitable for sedges (*Carex* spp.), which thrive where the silt is near the surface. They displace the reeds or bulrushes around the edge. By this time, however, the reeds or bulrushes have spread further towards the centre of the lake, migrating over the platform of silt accumulated by their underground organs.

As the silt builds up beneath the water, the sedges which occupy the original margin of the lake are replaced by sedge species which have an entirely different growth form. The rhizomes and dead leaf bases of species like the greater tussock sedge (*Carex paniculata*) form dense humps up to 1.5 m high and 1·m in diameter. Around their bases, silt accumulates. The tops, sides and bases of these humps are gradually invaded by those flowering plants which are characteristic of marshland.

Eventually the tops of these sedge mounds are invaded by alder (*Alnus glutinosa*), a tree which is capable of growing in moist soil. Since it has root nodules which contain symbiotic nitrogen-fixing organisms, alder increases the level of soil nitrogen. It forms a woodland around the edge of the lake, with marshy vegetation underneath. In this 'alder carr', silt continues to accumulate, and when the soil level has been raised above the water table, seedlings of oak may become established. Eventually, given hundreds of years, oak forest may replace the alder carr around the original margin of the lake.

All the time, the vegetation zones slowly march towards the middle. They form a series of contracting concentric rings with open water on the inside, followed in turn by reeds, sedges, hump-forming sedges and alder carr over marshland. Eventually the reeds or rushes meet in the middle. If the input of silt continues, the level of silt will rise until the reeds disappear. The same will happen eventually to the sedges, the hump-forming sedges, and the alder. Oak forest, perhaps with a stream running through it, may ultimately become established when the soil dries out. It may then persist for a long time as the climax vegetation.

FIGURE 53

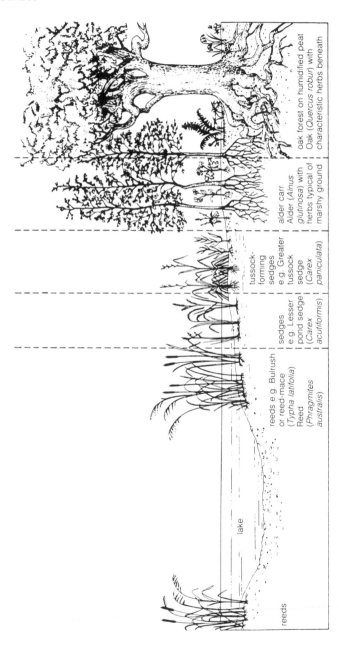

reeds

lake

reeds e.g. Bulrush or reed-mace (Typha latifolia) Reed (Phragmites australis)

sedges e.g. Lesser pond sedge (Carex acutiformis)

tussock-forming sedges e.g. Greater tussock sedge (Carex paniculata)

alder carr Alder (Alnus glutinosa) with herbs typical of marshy ground

oak forest on humidified peat Oak (Quercus robur) with characteristic herbs beneath

Succession from a lake to an oak forest. Silt accumulates around the bases of the plants, which raise the level and create conditions suitable for the invasion of other species. Evidence for the oak forest stage is scanty; peat bog may form instead.

The various stages in this process can be seen at present in the Shropshire and Cheshire meres, the Lake District, and several lakes in East Anglia. Research on the lakes in the Lake District suggest that the whole process from open water to oak forest may take about 10 000 years. The rate of succession, however, will vary considerably from lake to lake, since it depends on the volumes of the inflow and outflow streams, the area and depth of the lake, and the silt and nutrient content of the water.

What evidence is there that this successional process has occurred at all? Cylinders of soil, collected from beneath the alder carr and oak forest, contain a great deal of peat. The remains of plants accumulate in the silt because organic matter is only broken down slowly under anaerobic conditions. Sedges are also particularly resistant to decay. In the cylinders of soil, the remains of hump-forming sedges occur at the top. Beneath that is sedge peat, and at the base, the remains of reeds.

The dead plants at each level can be aged by 'radiocarbon dating'. This method relies on the fact that one of the radioactive isotopes of carbon, ^{14}C, decays with a half-life of about 5568 years. This isotope is present in the air in a small proportion of the carbon dioxide. Imagine that the $^{14}C/^{12}C$ ratio in the air was the same when one of these lakeside plants was photosynthesising as it is today. Originally, the carbon compounds deposited in its skeleton should have had the same $^{14}C/^{12}C$ ratio. Since then, however, the ^{14}C had decayed to ^{12}C at a known rate. Thus the present ratio of $^{14}C/^{12}C$ in these dead tissues provides a measure of the length of time which has elapsed since the tissue was deposited.

Because of this successional process in water, many lakes rich in nutrient ions in northern temperate regions seem doomed to extinction unless the inflow of silt is prevented. Even those lakes with a low nutrient content, which do not support much plant growth, undergo a natural enrichment process, called eutrophication, as material enters the lake from outside. This natural ageing process has been greatly hastened in many areas by nutrient enrichment from fertilisers, detergents and animal sewage. Thus, even if they begin life with a low level of nutrients, lakes are eventually likely to undergo plant succession.

Salt marsh succession

Between the tide marks, in shallow estuaries, the plants accumulate waterborne silt around their bases, just as they do in lakes. This causes a salt marsh to form.

In northern Europe, and North America, the zone just above the average low tide mark is colonised by plants like cordgrass (*Spartina* spp.) and glasswort. These species can tolerate long periods of immersion in sea water. Silt is caught around their stem bases and underground organs. This

FIGURE 54
Succession on a salt marsh. Silt accumulates around the bases of the first invaders, thus raising the level. The zones and species are those most likely in Britain.

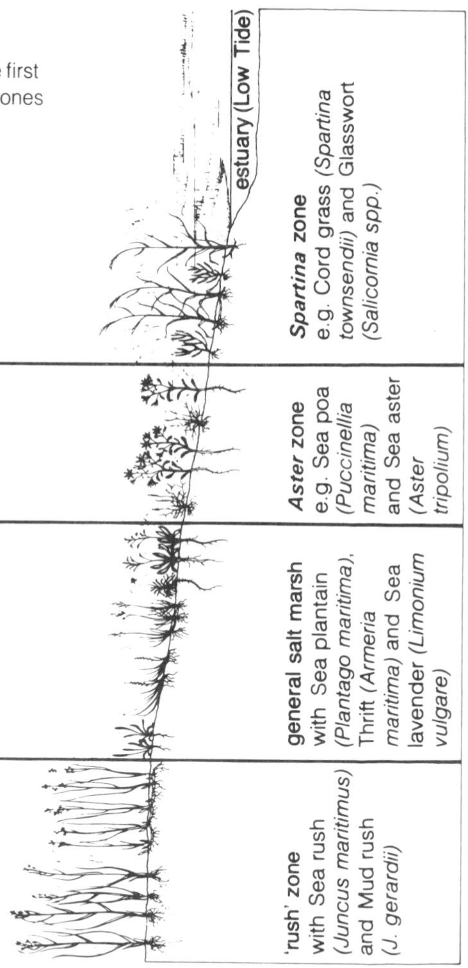

estuary (Low Tide)

Spartina zone e.g. Cord grass (Spartina townsendii) and Glasswort (Salicornia spp.)

Aster zone e.g. Sea poa (Puccinellia maritima) and Sea aster (Aster tripolium)

general salt marsh with Sea plantain (Plantago maritima), Thrift (Armeria maritima) and Sea lavender (Limonium vulgare)

'rush' zone with Sea rush (Juncus maritimus) and Mud rush (J. gerardii)

creates conditions which are more favourable to other plants, which then replace them. A highly predictable succession of species occurs (figure 54). At any one place, the level of the mud rises by a metre or so every one hundred years. As the silt level rises, the plants become submerged less and less frequently by the tides. Eventually, when only the highest spring tides can submerge the mud, this part of the marsh stops accumulating silt.

As the level of silt rises, each species moves slowly towards the centre of the estuary. A zonation develops as different plant species exploit different levels of the marsh. The reasons for this zonation, which reflects the succession, are still obscure. It seems to be the result of slightly different tolerances to waterlogging, submergence, salinity and nitrogen content of the soil, accentuated by competition between the species.

FIGURE 55
Succession across a sand dune. Sand accumulates around the bases of the plants. As succession proceeds the plants deposit more organic matter. This increases the capacity of the soil to hold water and nutrient ions.

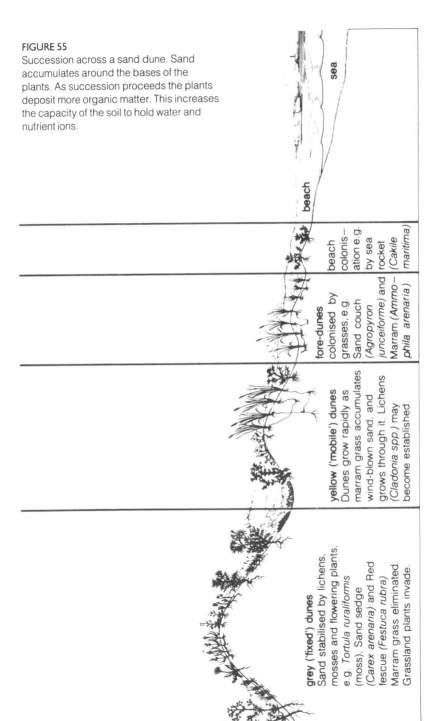

sea

beach

beach colonisation e.g. by sea rocket (*Cakile maritima*)

fore-dunes colonised by grasses, e.g. Sand couch (*Agropyron junceiforme*) and Marram (*Ammophila arenaria*)

yellow ('mobile') dunes Dunes grow rapidly as marram grass accumulates wind-blown sand, and grows through it. Lichens (*Cladonia spp.*) may become established.

grey ('fixed') dunes Sand stabilised by lichens, mosses and flowering plants, e.g. *Tortula ruraliformis* (moss), Sand sedge (*Carex arenaria*) and Red fescue (*Festuca rubra*). Marram grass eliminated. Grassland plants invade.

Sand dune succession

The mineral soil particles which are trapped by plants in sand dune succession are grains of sand, blown by the wind. The particles accumulate rapidly and may smother and kill the leaves of plants. On sand dunes, nutrient ions and water are often in short supply.

Sand on the landward side of beaches, well above the tide marks, is colonised first by certain plants (figure 55) which can tolerate high levels of wind-blown salt and a low soil water content. As sand is blown over the surface of the beach it accumulates around the plants, forming low mounds. These 'primary dunes' are colonised by various species of grass. These are the main dune builders. Marram grass (*Ammophila arenaria*), can grow up through a metre of sand a year. It accumulates sand around its leaves until it forms the characteristic tall 'yellow dunes' with its leaves projecting from the top. Its roots and rhizomes stabilise the sand.

As more dunes appear on the seaward side of the ridge, the dune which we are concerned with becomes more sheltered. Less sand is blown across its surface, the dune ceases to grow, and organic matter is deposited from marram grass and other plants. This organic matter glues together the soil particles, reduces sand movement, and holds water and nutrient ions. The sand is invaded first by small ephemerals, annuals and mosses, which speed up the rate at which organic matter accumulates and the soil is stabilised.

Eventually, conditions become suitable for the colonisation of grassland plants, which produce a great deal of organic matter and prevent the movement of sand over the surface. The water-holding capacity of the soil increases. Nutrient ions are trapped, both from rainfall and from nitrogen fixation by the root nodules of leguminous plants. Eventually, as the dune system moves out towards the sea, the dune ridges furthest from the sea become grasslands which, on sands rich in calcium carbonate, are usually grazed or mown. On acid sands, the grasslands are invaded by heather and then by willow, birch and pine.

The successions which have been mentioned so far are all rather idealised, because they assume no human intervention. In practice, hydroseres may be affected by dredging or the damming of streams, salt marshes are considerably influenced by grazing pressure and many sand dunes have been devastated by holidaymakers. The vegetation of a highly developed landscape like that of the British Isles tends to be a mosaic of communities in various stages of secondary succession.

SECONDARY SUCCESSION

Secondary successions take place on soils which have already been formed and colonised. They tend to occur much more rapidly than primary successions. This is because the rhizomes, seeds, spores or eggs of species already adapted to the site are still present, and because well developed soils already exist.

There are three main causes of secondary succession.

(a) Natural catastrophes which remove existing vegetation, like fires, hurricanes, flash floods, flooding, windblow or the spread of a parasitic fungus.

(b) Human destruction of climax communities, with the establishment of communities of an earlier successional stage. Man has cut and burnt down forests on a large scale, drained marshes, diverted rivers and streams, dug ditches, reclaimed land from the sea, and built roadside verges. Secondary successions can occur in all these places.

(c) Man maintains many communities at an early successional stage by regular ploughing, burning or grazing. If any of these pressures on farmland, grassland or heathland is relaxed, a secondary succession will occur.

A typical sort of secondary succession occurs on abandoned arable fields in lowland Britain. They support in turn arable weeds, a mixture of arable weeds and grasses, tall grassland dotted with shrubs like hawthorn, a mixed scrub, young oak/ash woodland and oak forest. The whole sequence may take 150–250 years. There is a field at Rothampsted Experimental Station in Hertfordshire which was abandoned in 1882 and has, after over a hundred years, reached the young oak wood stage (figure 56).

FIGURE 56

The Broadbalk Wilderness at Rothamsted Experimental Station, Hertfordshire. After the wheat harvest of 1882 an area of 0.1 ha was enclosed by fence, and has not been cultivated since. It is now woodland, mainly of hawthorn, oak, ash and sycamore (*Acer pseudoplatanus*). Some trees are over 18 m high. Over the period 1881–1964 the top 68.5 cm of soil has accumulated about 4.4 Mg ha^{-1} of nitrogen and 50 Mg ha^{-1} of organic carbon. (Rothamsted Experimental Station.)

The succession can be prevented if the site is grazed or burnt. The plant community which develops under this regime is not necessarily similar to one of the stages in the succession which would have occurred had the site been left to nature. The succession is 'deflected'. For example, a normal succession on arable chalky soils would be through herb-rich vegetation, which is then colonised by shrubs, to beech forest. Under sheep or rabbit grazing, however, grasses, not herbs, predominate. A short pasture develops which contains many palatable grass species.

Secondary succession is particularly important to man for the following reason. To prevent secondary succession from taking place in crops, in other words, to stop the entry of unwelcome plants and animals, a great deal of energy has to be expended (see p. 98).

Many climax communities appear to be mosaics of areas in different stages of secondary succession. In areas which are subject to periodic climatic catastrophes, such as floods, fires or hurricanes, the vegetation may be devastated at intervals and replaced by trees which form an even-aged group. Conversely, in climatically stable areas, dead trees may create gaps, which are exploited by the existing saplings or new seedlings which take part in a mini-succession. In fact the diverse species composition of many forests may be maintained by 'reciprocal replacement', in which the saplings of each species stand more chance of becoming established under trees of other species.

Partly because of primary and secondary succession, the surface of an intensively cultivated country like Britain supports many different types of community. There are arable fields, artificial grassland, grasslands of many different kinds, hedges, roadside verges, copses, scrub, rivers, marshes, fens, bogs, woodlands, tundra, plantations of alien trees and so on. This sort of variety is difficult to analyse, particularly since each type of community contains a different mixture of interacting plant and animal species. One way to reduce this complex situation to manageable proportions is to think of each community as part of a different ecosystem.

6 Ecosystems

Ecologists frequently study small areas and concentrate upon food webs, trophic levels, energy flow or nutrient cycling (see chapter 1). In that case they are studying aspects of ecosystems. An ecosystem is a community of organisms, interacting with one another and their inanimate environment (e.g. climate, soil) to form a *more or less* balanced, self-sufficient unit with its own characteristic pattern of energy flow and nutrient cycling.

The idea of an ecosystem which exchanges energy and compounds with its environment is depicted in figure 57. It is important to emphasise that an ecosystem does not merely consist of a group of organisms. It also includes the non-living environment in which the organisms live. The word 'ecosystem' often implies that it is self supporting, in the sense that the green plants take up from the environment most of the energy, ions and compounds which the other organisms need.

The bare essentials of an ecosystem are producers, decomposers, and inorganic compounds such as water, carbon dioxide and nutrient ions. The producers, the green plants, extract these inorganic compounds from their environment. The decomposers, the bacteria and fungi, return these materials to the environment. Most ecosystems exchange carbon compounds, nutrient ions, gases and water with others. This exchange is often unequal. Many inland lakes, for example, receive carbon compounds and nutrient ions from other ecosystems via rivers and streams.

An ecosystem can be of any size. It can range over several biomes. It might cover only a tiny area, like a small pond or a puddle. When ecologists study ecosystems, they usually study small areas which are representative of wide ranging types of community.

There are four main reasons why the ecosystem concept is valuable.

(a) When organisms are grouped into trophic levels, the immensely complex situation in nature is reduced to a small number of variables. This simplified picture can be easily grasped by the human brain and facilitates analysis.

(b) The concept focuses attention on energy flow and nutrient cycling

FIGURE 57

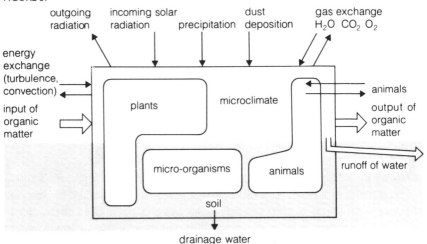

Diagrammatic representation of an ecosystem (framed) exchanging energy and compounds with its environment. (Adapted from H. Walter (1973). *Vegetation of the Earth in Relation to Climate and the Ecophysiological Conditions.* Springer–Verlag, New York.)

FIGURE 58

Simplified representation of energy flow and nutrient cycling in an ecosystem.

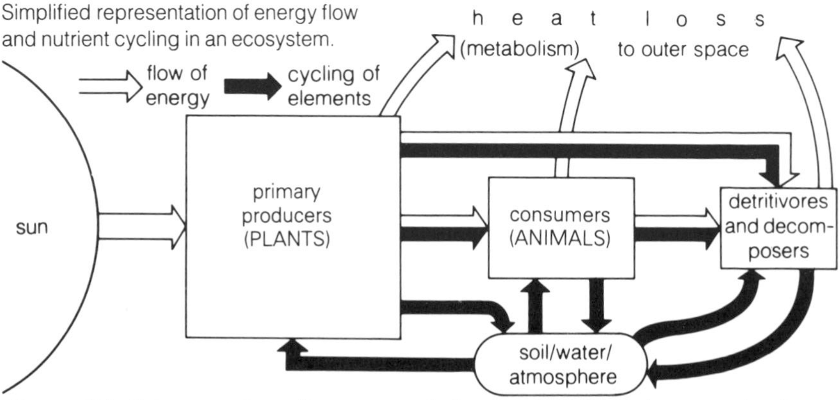

(figure 58). The trapping of energy and the uptake of nutrients are basic activities of all living organisms.

(c) Once the organisms in an ecosystem have been placed in trophic levels, the rate at which each trophic level accumulates energy or nutrients can be established. Its rate of plant production, animal production, or nutrient cycling can then be compared with those of other ecosystems. For example, the same units (kJ energy trapped $m^{-2}year^{-1}$) can be used to compare the plant productivities of two ecosystems which contain different sets of species (figure 59). Such comparisons highlight the environmental factors which are limiting the efficiency of primary and secondary production throughout the world. They also allow us to assess the relative potential of different ecosystems for food production.

FIGURE 59

Biome	Net primary production per unit area $(kJ\ m^{-2}\ year^{-1})$ on average
Extreme desert, rock and ice	60
Desert scrub	1320
Subsistence agriculture	1528
Open ocean	2420
Arctic and alpine tundra and heathland	2650
Continental shelf of oceans	6620
Temperate grasslands	9240
Lakes, rivers and streams	9450
Temperate woodland and scrub	11340
Industrialised agriculture	12290
Boreal coniferous forest	13100
Tropical savannah	13440
Temperate deciduous forest	22210
Tropical swamps and marshes	35280
Tropical estuaries and attached algae	35280
Moist tropical evergreen forests	36160

Net primary production in the major biomes of the biosphere.(After R. E. Ricklefs (1973), R. H. Whittaker (1975) and H. Leith (1975)). Primary productivity of the major vegetation units of the world. *In*: H. Leith and R. H. Whittaker (eds). *Primary Productivity of the Biosphere. Ecological Studies 14.* (Springer–Verlag, Berlin 203–15.)

(d) Ecosystems, including crop ecosystems, provide the life-support systems for humans and other animals. The more we know about how they function, and how they behave in response to disturbances like harvesting and pollution, the better. In the end we should be able to accurately predict the long-term consequences of any unusual stresses to which an ecosystem might be subjected.

Following a discussion of trophic levels and food webs, much of the rest of this chapter is devoted to energy flow. Primary production, secondary production, energy flow from one trophic level to the next, and ecological pyramids are dealt with in turn. Then the patterns of energy flow in different ecosystems are compared. After a brief discussion of the major nutrient cycles, and of the relationship between energy flow and nutrient cycling, the chapter ends with several examples of human impacts on ecosystems and the need to understand them.

TROPHIC LEVELS AND FOOD WEBS

Ecosystem ecologists usually wish to work out in numerical terms how much energy and nutrient flow occurs between one trophic level and another. To make this possible, the main organisms in the ecosystem must each be assigned to a trophic level. At first sight this might seem easy. Most plants, for instance, are primary producers. Yet in practice, many organisms occupy more than one trophic level.

Consider yourself, for example. You can be a primary, secondary and tertiary consumer at the same time. Insect-eating plants, like Venus' flytraps (*Dionaea muscipula*), act simultaneously as primary producers and as primary or secondary consumers. Many omnivores, such as voles, eat not only plants but herbivores and detritivores as well. Into which trophic level should voles be put? In practice, if 60% of the mass of a vole's diet consists of plant matter and 40% consists of primary consumers, then 60% of its energy or nutrient intake would be assigned to the primary consumer trophic level, and 40% to the secondary consumer level. Only in this way can the mind-boggling complexity of a natural ecosystem be reduced to its essential details.

Our knowledge of most organisms is much more basic than this. We do not know what they eat, let alone the relative proportions in the diet of the various types of foodstuff. In order to place a species in a trophic level, its diet must be investigated.

Four main methods have been used to find out what organisms eat.

(**a**) Observation is sometimes useful, for example in the case of leaf eating insects. Predators can often be seen eating their prey, but one can never be certain that a predator has been seen in the act of eating *all* its usual prey items.

(**b**) In food preference experiments the animal is provided with known masses of a range of possible food items. After some time, these foods are weighed again to find out which were preferred. Unfortunately this reveals little about what the predator eats in nature, because the frequency with which it encounters and catches each prey item in nature is unknown.

(**c**) The faeces, gut contents or regurgitated pellets of animals are analysed for the identifiable remains of their prey. The exoskeletons, bones or leaf epidermises of animals and plants often enter the faeces almost unchanged. Unfortunately, soft-bodied plants and animals are easily digested and may be underestimated or missed. To detect the presence of the juices of such organisms in the gut the precipitin test can be used. An extract of a suspected prey species is injected into a rabbit, which makes specific proteins (antibodies) against the alien proteins. The rabbit blood serum containing the antibodies is combined with juice from the gut of the suspected predator. The formation of a precipitate shows that an antibody–antigen reaction has occurred and that the predator must have recently eaten the suspected prey species.

(d) Radioactive isotopes can be used to trace food webs. The plants of a single species are labelled, for instance with ^{32}P. Samples of the animals nearby are captured at regular intervals of time. All the animals which fed on the plants directly or indirectly will be radioactively labelled. The herbivores will reach their peak of radioactivity soon after the experiment has started. The top carnivores, however, will accumulate the isotope much more slowly.

Once the diet of the major species in the ecosystem has been determined, and the organisms have been assigned to trophic levels, food webs can be drawn. Then the patterns of energy flow and nutrient cycling can be dimly discerned. Energy flow, discussed in outline in the first chapter, will be considered first.

ENERGY FLOW

The energy source for ecosystems is short wave radiation from the sun. Yet energy, once created, cannot be destroyed. The same energy, mainly as long wave radiation, is lost from the ecosystem to outer space, at the same average rate as the ecosystem absorbs it.

Most of the energy is not used to power living organisms. It is reflected from the soil, water or vegetation, or absorbed by them and re-radiated as heat.

A small proportion of the incoming energy, however, is trapped by photosynthetic pigments and some of this enters the bond energy of molecules like glucose. The plants, animals and micro-organisms in the ecosystem depend on the energy in this glucose to drive their chemical reactions, with the exception of the chemautotrophic bacteria, like *Nitrobacter*, which derive their energy from the oxidation of inorganic compounds. As a result of all these reactions, the energy which was trapped in the glucose molecules is ultimately released to space as heat (= long wave radiation).

In this section primary production, secondary production and decomposition are dealt with first. Then energy flow through different ecosystems is compared.

Primary production

The net primary production of an ecosystem (NPP) is the rate at which the plants accumulate dry mass, and is measured in units of $kg\,m^{-2}\,year^{-1}$. This growing store of energy and dry mass is potential food for the animals in the ecosystem. Alternatively, the plant matter can be exploited by humans, for instance, in crop ecosystems or forests. It is important for us to understand the processes which limit the efficiency of net primary production if we are to increase our production of food and fuel.

The net primary production is the difference between the rate at which the plants photosynthesise (Gross Primary Production–GPP) and the rate at which they respire (R). In other words, the glucose molecules produced in photosynthesis have two main fates. Some of them are used to provide the energy for maintenance, growth and reproduction, and their energy is lost from the plant as heat during processes such as respiration (R). The rest are deposited in various forms in and around cells and represent stored dry mass (NPP = GPP − R).

Over the whole of the earth's surface, total photosynthesis (GPP) traps 3×10^{21} J of sunlight energy each year. The total mass of stored compounds (NPP) produced each year as a result of photosynthesis (GPP) minus respiration (R) is 200 000 Tg, about three hundred times the world's annual production of steel. Yet of the total sunlight energy reaching the earth's surface, only 0.1% is stored in plant mass (NPP) in this way. Why are plants apparently so inefficient at trapping energy from the sun?

The sun is an incandescent ball of gases with its outer surface about 6000°C. The continuous fusion of nuclei of hydrogen atoms to produce helium nuclei produces electromagnetic radiations of wavelength 300–10 000 μm which travel about 150 Gm through space before they encounter the earth's atmosphere.

Energy from the sun reaches the outside surface of the atmosphere of the earth at an average rate of $1.4 \text{ kJ m}^{-2} \text{ s}^{-1}$, called the solar flux. Of this, thirty to forty per cent, including most of the infrared (wavelengths >700 nm) is reflected away from the earth's surface by clouds or dust. Another ten to twenty per cent is absorbed by gases, water vapour or dust in the atmosphere and is converted to heat energy. This includes the dangerous ultraviolet wavelengths (wavelengths <400 nm). They are absorbed by the ozone shield which surrounds the upper atmosphere at an altitude of 20–50 km. The rest of the radiation reaches ground level, where there is thus an average daytime intensity of solar radiation of $0.8–1.0 \text{ kW m}^{-2}$. Let us call this 1000 units of energy, and trace what happens to it as it is intercepted by a typical crop plant (figure 60 p. 93).

Of the energy reaching ground level on a clear day, about 10% is in the ultraviolet, and 45% is in the infrared. Only 45% is in the visible wavelengths (400–700 nm) which could be used by green plants in photosynthesis.

Much of this energy is reflected from plant leaves, passes through them, or is absorbed by earth or water without having hit plant tissue. Most of the energy absorbed by the stems and leaves of plants is used to evaporate water. Only a small proportion is absorbed by pigments in the chloroplasts. Under certain conditions, for instance when the plant is under water stress or the light intensity is high, some of this absorbed energy may be wasted

FIGURE 60

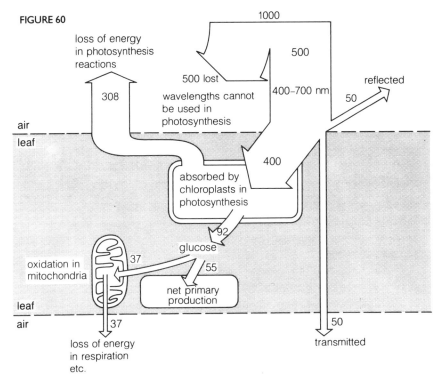

The average efficiency with which total solar radiation is trapped in net primary production in a productive crop plant (5.5%). Imagine that the plant grows at sea level and receives 1000 units of light energy per unit time. Much of the 'wasted' energy is used to evaporate water. (After D. O. Hall (1979). *Biologist* **26**, 17.)

because the plant is already light-saturated. The rest will be used in photosynthesis.

Of the energy in the solar radiation which hits the pigments and is used in photosynthesis, only about a quarter ends up in the bond energy of molecules like glucose. The rest is lost in various phases of the light and dark reactions. Much of the energy in the glucose is released in cellular respiration and other reactions necessary for the plant to maintain itself and grow. The rest represents the net primary production.

In fact the figure of 5.5% quoted here represents the maximum conversion efficiency achieved by a crop over a short period of time. Yet most fields are uncultivated for part of the year, and when the crop plants are small they do not intercept much of the incoming light. Over the whole year, the proportion of the total incoming energy which is converted into energy in stored carbon compounds (NPP) is typically 0.5–1.5% in temperate crops like wheat and 0.5–2.5% in tropical crops like sugar cane.

Nor does this small proportion represent the energy available to man or herbivores. In the case of grain crops, for example, the grain is harvested but most of the accumulated dry matter is discarded. Similarly, although most herbivores seem to be born into the midst of plenty, most are only able to eat particular parts of certain plant species at a single time of year.

The efficiency with which sunlight energy is incorporated in stored glucose molecules (NPP) is also low in semi-natural ecosystems. About 0.4% of the total incoming radiation ends up in net primary production in grasslands, 0.5–1% in forests and 0.01% in the sea. Since the sea makes up nearly three-quarters of the earth's surface, the worldwide efficiency of energy conversion is rather low.

The factors which influence net primary production

The total plant growth each year in an ecosystem depends mainly on the intensity of solar radiation, the temperature, and the water and nutrient supply.

The maximum net primary production per unit area in semi-natural ecosystems occurs in tropical rain forests and tropical estuaries (figure 59). These both have high light intensities, high temperatures, plant growth all year round, and plenty of water. In other tropical and semi-tropical biomes, like deserts and savannahs, plant growth is limited by water shortage for at least part of the year. In general, the further from the equator one goes, the lower the light intensity, the lower the average temperature, the more seasonal the environment, the shorter the growing season and the lower the net primary production (figure 52 p. 77).

Primary production in fresh water, in the sea and on land is frequently limited by a shortage of nutrients. In temperate lakes and seas, a deficiency of nitrogen, phosphorus or silicon in the upper layers of the water has been deemed responsible for the very low primary production in mid-summer. This is just the time of year when the light intensity and temperature are highest. Eutrophicated lakes have a much greater net primary production of algae during summer than unfertilised lakes. On land, the yields of plants in most ecosystems can be increased by the application of nitrates or phosphates. Fertilisers are applied to grain crops like wheat to increase their rates of growth. This evidence suggests that a lack of nutrient elements often limits the growth rate of plants.

Primary production in the sea, and in crop ecosystems on land, affects our food supply. The factors limiting primary production in these ecosystems are therefore of particular interest. They will be discussed briefly in turn.

Primary production in temperate lakes and seas

In temperate lakes and seas, there are two peaks of phytoplankton dry mass each year (figure 61). In spring the rapidly increasing temperature and light intensity stimulate a population explosion of phytoplankton. At about the same time the upper layers of water warm up and become less dense. They form a warm light layer, the epilimnion, which is 10–300 m deep in deep seas and lakes. This floats on a cold dark layer, the hypolimnion, which remains close to 4°C. Between them is a transition zone called the thermocline.

In spring and summer there is very little exchange of water between the epilimnion above and the hypolimnion below. The result is that the phytoplankton deplete the epilimnion of nutrients. The nutrient elements like nitrogen and phosphorus fall, within dead phytoplankton, to the bottom. There they are released by the activity of bacteria, but they are not recirculated to the epilimnion.

The decline in the numbers of phytoplankton in late spring may partly be caused by the reduction in the nutrient content of the upper layers of the water. Shortage of nutrients may also limit the primary production of the phytoplankton during the sumer, maintaining it at a low level. Circumstantial evidence for this is that primary production in the sea (and therefore the secondary production of fish) is concentrated in areas off Antarctica and Peru where cold currents rich in nutrients well up against continents and bring nutrients to the surface (figure 62).

FIGURE 61

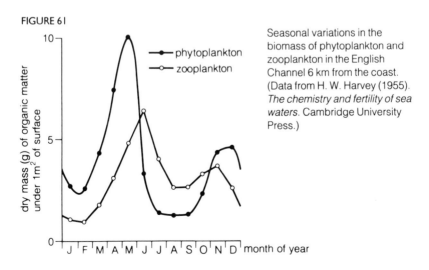

— phytoplankton
—○— zooplankton

dry mass (g) of organic matter under 1m² of surface

month of year

Seasonal variations in the biomass of phytoplankton and zooplankton in the English Channel 6 km from the coast. (Data from H. W. Harvey (1955). *The chemistry and fertility of sea waters.* Cambridge University Press.)

FIGURE 62

Zone	Percentage of ocean area	Average net primary production $(kJ\,m^{-2}\,year^{-1})$	Average number of trophic levels	Average secondary production of fish (Gg)
Open ocean	90	2100	5	160
Coastal	9.9	4200	3	120 000
Coastal upwelling areas	0.1	12 550	1.5	120 000

The primary and secondary production of different zones of the sea. (J. H. Ryther (1969). Photosynthesis and fish production in the sea. *Science* **166**, 72–6.)

Another reason why the biomass of phytoplankton declines in late spring could be that they are grazed by animal plankton (= zooplankton). The zooplankton (figure 63) such as small crustaceans, increase in numbers just as the phytoplankton are declining. These animals probably keep the numbers and mass of phytoplankton at a low level throughout the summer. In lakes, at least, the spring 'bloom' of phytoplankton consist of edible species but the autumn peak contains mainly unpalatable ones. This is what might be expected if the zooplankton graze the edible phytoplankton all summer.

The autumn burst of net primary production occurs when the thermocline is overturned by cold stormy weather. The epilimnion and hypolimnion mix. The nutrients reach the upper layers of the water, and this occurs at a time when the light intensity and water temperature are still high enough to support rapid algal photosynthesis. The numbers of phytoplankton rapidly increase, but decline abruptly when the cold weather of winter arrives.

FIGURE 63

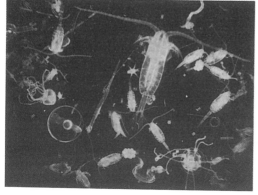

Zooplankton. Living marine plankton, magnified about eight times. The photograph includes several species of copepods (the largest is *Calanus helgolandicus*), a fish egg, an arrow-worm (*Sagitta*), and two anthomedusae. Also present, very small, are diatoms, some in chain form. (Courtesy Douglas P. Wilson).

Primary production in agricultural crops

Large-scale agricultural production depends mainly on the net primary production of certain grasses, such as oats, barley, rye, wheat, paspalum, sorghum, millet or maize. Net primary production in these grasses may be limited by a shortage of light, water, carbon dioxide, or nutrient ions. A superabundance of herbivores may also reduce crop growth. If net primary production is to be increased, some of these constraints must be removed.

Crop yields often improve when pests are eliminated or water or nutrient ions are supplied. Industrial agriculture now depends on insecticides, fungicides and herbicides to eliminate pests, and on artificial fertilisers to increase crop growth. The water supply can be improved by irrigation or by maintaining a high humus content in the soil. If these measures are employed the crop grass is being grown in an artificially improved environment. On the other hand, the light intensity and the carbon dioxide concentration cannot easily be changed. They can only be increased artificially on the scale of a greenhouse.

Better growing conditions thus produce higher yields. The other side of the coin is that new crop varieties have been developed to exploit these better growing conditions to the full. There has been unconscious and conscious artificial selection for varieties of grass which produce the greatest mass of useful dry matter per unit area.

What characteristics should we expect of an ideal crop plant? Its seeds should germinate rapidly, and the young plants should quickly cover the whole of the ground surface with leaves. Eventually, a canopy of leaves should capture most of the solar radiation. The crop plant should be able to respond to an increasing light intensity with an increased rate of photosynthesis, right up to the intensity of full sunlight, so that none of the sun energy is wasted. It should be resistant to fungal attack, and robust enough to survive buffeting by high winds. It should be capable of rapid growth. Fertilisers should be efficiently absorbed from the soil and the nutrient ions should stimulate the production of new cells at the meristems. Losses of glucose through cellular respiration should be as small as possible. Finally, the plant should transfer a high proportion of its dry mass into the parts that we harvest.

All these characteristics have been selected for by plant breeders. Many different strains of crop plants have been produced, each particularly suited to a different range of environmental conditions.

One major factor which determines the environment to which a plant is best suited is whether it photosynthesises by the Calvin cycle (C_3) or whether it uses the Hatch and Slack (C_4) pathway too. The leaves of C_4 plants like maize and sugar cane have a considerable affinity for carbon dioxide, and can photosynthesise rapidly at high temperatures and light intensities. At high temperatures, C_3 plants have increased respiration rates through 'photorespiration', which only occurs in the light and wastes a great deal of energy. C_3 plants are also light saturated at much lower light intensities than C_4 plants. Thus in tropical or semi-tropical conditions, the net primary production of a C_4 crop is usually much higher than a C_3 crop.

In primitive agricultural systems the energy to clear the ground, spread the seed, remove weeds and harvest the crop is expended by humans. Some energy from wood or peat is expended to make tools and simple machinery, but the energy output in grain is more than twenty times the energy input.

In industrialised agriculture the net primary production of grasses can be eight or ten times higher than in subsistence agriculture. Yet these high yields are only achieved at the expense of energy released in the burning of fossil fuels. Coal, gas, peat and oil are used to build and drive farm machinery, and to produce and apply pesticides and fertilisers (figure 64). In the industrialised countries 2 kJ of fossil fuel energy are expended to produce each kJ in grain. In the extreme case, greenhouse crop production, about 100 kJ of fossil fuel energy is used in trapping 2 kJ of food energy. Howard Odum has called crops produced like this 'potatoes made of oil'. As fossil fuels become scarcer and more expensive, it may become difficult to sustain these high crop yields.

Two possible remedies are at hand. Firstly it might be possible to induce crop plants to fix atmospheric nitrogen. This would reduce the dependence of agriculture on nitrogenous fertilisers produced by the Haber–Bosch process, which expends a great deal of energy. Secondly, plants themselves can be grown for fuel, reducing our reliance on fossil fuels. In Brazil, for example, ethanol is being produced from sugar cane and cassava, and is added to petrol on a large scale.

Net primary production is important because it provides the food for consumers. We shall now turn our attention to the accumulation of energy by consumers.

Secondary production

Secondary production refers to the rate at which consumers in food chains accumulate dry mass (= biomass) or energy. Consider the animals at a particular trophic level. Their secondary production is the dry mass laid down, or the energy in the dry mass laid down, per unit area per unit time.

The production of the animals at a particular trophic level must be far less than the production of the organisms on which they feed. In the first place, animals do not eat all the mass available to them. Caterpillars eating the leaves on a tree, for example, rarely defoliate the whole tree. In the second place, only a small proportion of the energy which an animal eats becomes deposited in the mass of its tissues.

FIGURE 64

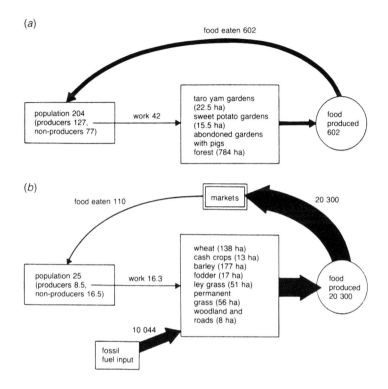

(a)

(b)

Energy flow in subsistence and industrialised agriculture. The figures represent energy flows in GJ per year.
(a) Flow through the shifting agricultural system of the Tsembanga Maring tribe, in the mountains of New Guinea. Producers numbering 127 produce 602 GJ of food.
(b) Energy flow through a mixed farm at West Lavington, Wiltshire, in 1971. Producers numbering 8.5 produce 20 300 GJ of food, aided by a massive subsidy from fossil fuels. (Modified from T. P. Bayliss-Smith (1982). *The Ecology of Agricultural Systems*. Cambridge University Press.)

Let us trace the fate of the energy in the food taken by a stick insect, a primary consumer. Some of its total food intake will pass along its gut and will be egested at the anus without having entered the caterpillar's body. The rest will be absorbed across the wall of the gut (= assimilated).

Of the assimilated energy, some will be used in cellular respiration to provide the stick insect with energy for movement and the manufacture of new chemical compounds. Some will be removed in the nitrogenous waste which is excreted, known as urine in many animals. The rest will be stored in the dry mass of new tissues as secondary production.

This relationship can be expressed in terms of an equation written in units of dry mass or energy:

$$FE = E + R + U + P$$

where FE is the food eaten, E is egesta, R is the expenditure on cellular respiration, U is urine and P is secondary production. Rearranging this equation,

$$P = FE - E - R - U$$

The secondary production, P, represents the mass or energy available to the organisms at the next trophic level. The relationship between dry mass and energy can be determined by burning dry tissue or egesta in a 'bomb calorimeter'. A known mass of organic matter is burnt in pure oxygen in an insulated container, and from the temperature which the apparatus reaches the energy content of the sample can be calculated.

Some figures have been worked out for a bullock grazing on a meadow (figure 65). The net primary production of the vegetation is about 21 135 kJ m^{-2} year^{-1}. The bullock only eats the leaves of the most palatable plants, and much of the energy trapped in photosynthesis must enter dead leaves or roots, which the bullocks do not eat. He takes in by mouth about 3053 kJ m^{-2} year^{-1}.

Well over half of this (62.5%) passes straight through the gut and is egested as cow-pat or methane gas. The methane produced in a bullock's gut by mutualistic protozoa and bacteria, about 60 l h^{-1}, is enough to light a

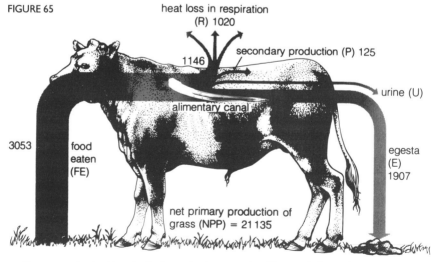

FIGURE 65

heat loss in respiration (R) 1020

1146

secondary production (P) 125

urine (U)

alimentary canal

3053 food eaten (FE)

egesta (E) 1907

net primary production of grass (NPP) = 21 135

The energy budget for a bullock grazing in a meadow. The symbols are those used in the text. The figures are energy flows in kJ m^{-2} year^{-1}. (*Nuffield O-level text III. The maintenance of the organism.* Longmans.)

small house continuously! The other 37.5%, amounting to $1146 \, kJ \, m^{-2}$ $year^{-1}$, is assimilated, which in this case means taken through the gut wall into the blood stream or the lymphatic system.

Most of the assimilated energy (89%, $1020 \, kJ \, m^{-2} year^{-1}$) is expended in cellular respiration to keep the animal alive and kicking. A bullock has a particularly high respiration rate because it is an endotherm with a constant body temperature of about 39°C. The other 11%, about $125 \, kJ \, m^{-2} year^{-1}$, is incorporated in new cells, of which many will be available to humans as food when the animal is slaughtered.

This means that only 0.6% of the net primary production of the grass has been incorporated into the secondary production of the bullock. It is obvious that many more humans could be supported on the primary production of grass than the secondary production of beef cattle. Not all consumers, however, are such poor energy converters as bullocks.

The efficiency with which food is converted into biomass
The efficiency with which consumers convert the food which they take in by mouth to biomass depends on two main factors. Firstly, is the consumer an endotherm or an ectotherm? Secondly, how much of its diet can it assimilate? The influence of these factors is quite clear from figure 66.

			Taken in		Net pro-	
	Feeding	Egested	through	Respired	duction	R/P
Animal	preference	(E)	gut wall	(R)	(P)	ratio
Grasshopper	Herbivore	63	37	24	13	1.8
Caterpillar	Herbivore	59	41	17.5	23.5	0.7
Wolf spider	Carnivore	8.2	91.8	57	27	2.1
Perch	Carnivore	16.5	83.5	61	22.5	2.7
Owl	Carnivore	15	85	85	0	high
Elephant	Herbivore	66	33	32	1	32
Cow	Herbivore	60	40	39	1	39

FIGURE 66 — *Arbitrary units of energy. Imagine that each organism has eaten 100 units*

The efficiency of assimilation and secondary production in selected animal species. The horizontal line half way down the table separates the ectotherms (poikilotherms) above from the endotherms (homiotherms) below. (Mainly from Open University S323 course, (1974) *Unit 3. Consumers in ecosystems.* Open University Press.)

The final column in the table illustrates that in endotherms a much higher proportion of the assimilated energy is devoted to respiration than in ectotherms. The respiratory costs of maintaining a constant body temperature of 35–42°C are massive, particularly in old climates or cold seasons of the year. At the temperatures which prevail in temperate regions, usually between 5 and 25°C, endotherms have to warm themselves up by respiration but ectotherms do not. The energy which endotherms 'save' can be devoted to the production of new cells.

The second column of the table shows that carnivores assimilate a much higher proportion of their diets than herbivores. Carnivores such as secondary and tertiary consumers have high protein diets which are easily digested and absorbed. They assimilate over 80% of the energy in their diets and egest less than 20%. In contrast, much of the energy in a herbivore's food is present in the glucose in cellulose cell walls. Slugs and snails have cellulase enzymes. Ruminants and termites have mutualists which make some of the energy in the cellulose available to them. Nevertheless, they assimilate about 40% of the energy in their food, and egest the other 60%.

These figures for individual animals suggest that the transfer of energy between trophic levels is rather inefficient. This is confirmed by energy budgets for whole ecosystems, like that for Silver Springs, a large stream in Florida (figure 67). At each trophic level, energy is lost through death, egestion and respiration. The remaining energy is deposited in biomass and is available for organisms at the next trophic level to use. The result is that each trophic level takes up energy more slowly than the previous one.

This suggests one measure of the efficiency of energy transfer between trophic levels. The gross ecological efficiency (GEE) is the rate at which energy enters one trophic level, e.g. the secondary consumers, divided by the rate at which energy enters the previous trophic level, e.g. the primary consumers. For the Silver Springs data, the energy entering the secondary consumers was $1609 \, kJ \, m^{-2} \, year^{-1}$ and the energy entering the primary consumers was $14\,146 \, kJ \, m^{-2} \, year^{-1}$. The GEE is therefore 1609 divided by 14 146, which is 0.114 or 11.4%.

Slobodkin has pointed out that the energy taken in by the animals at one trophic level is *about* 10% of the energy taken in by the animals at the previous trophic level. In other words, GEE = 10%. This is the case, at least, in artificial food chains set up in the laboratory. It is a useful generalisation to bear in mind and is sometimes called the '10% rule'. Three points about this ratio need emphasis. Firstly, it refers to the food taken in by mouth by animals, FE in terms of the equation of p. 100. Secondly, it applies to the transfer of energy between primary and secondary consumers, secondary and tertiary consumers and so on. It does *not* apply to energy flow between plants and the primary consumers which eat them. Thirdly, it applies to whole trophic levels and not to individual species.

One important aspect of energy flow has yet to be discussed. Dead organisms and their egesta often contain a great deal of energy and are attacked by the detritivores and decomposers.

FIGURE 67

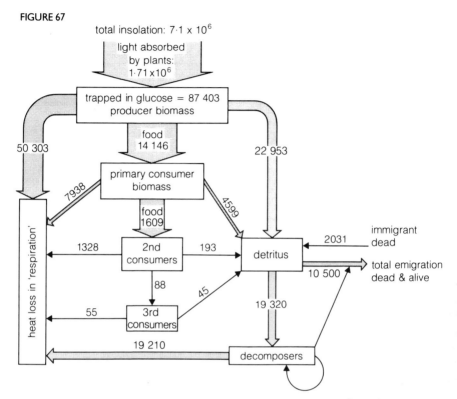

Energy flow through Silver Springs, Florida in 1953–4, in units of kJm^{-2} year^{-1}. Sizes of squares and widths of arrows are *not* proportional to biomass or energy flow. (H. T. Odum (1957). Trophic structure and productivity of Silver Springs, Florida. *Ecological Monographs* **27**, 55–112.)

Decomposition

Primary production and secondary production take place mainly in the upper layers of an ecosystem. In the sea, for example, the light is trapped by algae which are most abundant in the top 10 m of water. In a forest, the light is intercepted by chloroplasts several metres above the ground. The herbivores are most abundant in these zones because their food occurs there. Decomposition, on the other hand, occurs lower down, on or near the surface of a substrate like the sea bed or the soil. Gravity ensures that dead organic matter will reach the detritivores, bacteria and fungi which are established there.

The pattern of decomposition for the leaf litter in a forest is shown in figure 68. The dead leaves are first attacked by yeasts, other fungi, earthworms, springtails and mites. They use some of the energy to survive, and increase the surface area of detritus on which fungi and bacteria can act.

FIGURE 68

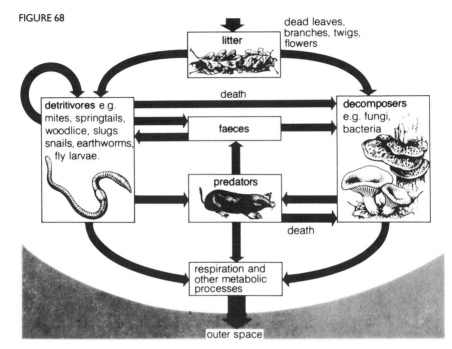

Energy flow in the decomposition of the leaf litter in a forest. Arrows point in the direction of energy flow. Organic compounds released to the soil from each box form humus. Nutrient ions are released to the soil from each box.

The fungi and bacteria secrete enzymes which break down the organic compounds in the detritus. In the process they break down dead detritivores, their egesta, and dead fungi and bacteria. They absorb from the remains of the cells the soluble organic monomers and the nutrient ions which they need to survive. The nutrient ions, essential for plant growth, are eventually released to the soil. Some organic compounds remain in the soil as humus.

In the end, the energy which reaches the detritivores and decomposers in the lower layers of an ecosystem is released to outer space as long wave radiation as a result of cellular respiration. This is shown in figure 67 for the Silver Springs data.

It is possible to express energy flow through the producers, consumers and decomposers in a single visual manner so that energy flow through one ecosystem can be compared with energy flow through another. There are two ways of doing this. One method is to draw 'ecological pyramids'. Another is to draw diagrams representing the energy budgets for whole ecosystems.

ECOLOGICAL PYRAMIDS

Ecological pyramids convey rapid visual impressions of the relative importance of the various trophic levels in an ecosystem. Three sorts of pyramids can be drawn, pyramids of numbers, biomass and energy.

To produce a pyramid of numbers, the number of organisms per unit area at each trophic level is counted. Then a bar diagram is drawn (figure 69). The numbers are expressed on a horizontal axis, so that the length of each horizontal bar represents the number of organisms at that particular trophic level. The bar at the base represents the number of producers, above that the number of primary consumers, above that the number of secondary consumers, and so on. Detritivores, decomposers and parasites are placed at the side.

This type of pyramid is of limited value for comparing ecosystems. Firstly, it is difficult to count the numbers of grasses or algae per unit area. Secondly, it does not take into account the relative sizes of organisms. A bacterium would count the same as a whale which weighed 130 Mg. This anomaly sometimes makes the pyramids inverted. In a forest, for example, the number of trees is small compared with the number of herbivorous insects (figure 69). Thirdly, the shape of the pyramid differs at different times of year.

Pyramids of biomass would take account of the difference in size between organisms. In such pyramids the horizontal axis is not the number of organisms at a particular trophic level, but their total dry mass (figure 70). Unfortunately comparison between the masses of different trophic levels in the same ecosystem, or the same trophic level in different ecosystems, are not very meaningful. This is because in each case the biomass has accumulated over a different period of time. The trees in a forest, for example, have stored their biomass over a hundred years or so. The equivalent primary producers in a lake, the algae, only contain carbon compounds which have been made over the previous few days. Furthermore, in the case of algae, their biomass fluctuates considerably throughout the year and so the shape of the 'pyramid' would depend upon the time of year they were sampled.

The pyramids of biomass for marine and freshwater ecosystems are sometimes inverted (figure 70). How can the zooplankton and fish weigh more than the phytoplankton when they both depend on the phytoplankton for food?

The answer is that the phytoplankton have a more rapid turn-over rate. An alga might have a life expectation of two days, whilst a small crustacean feeding on the alga might live for two weeks and a fish might last for two years. If one added up the total biomass produced by the algae over the whole year, it would exceed the total biomass gained by the primary con-

FIGURE 69

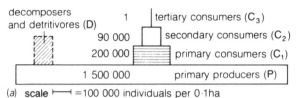

decomposers
and detritivores (D)   1   tertiary consumers (C₃)

90 000   secondary consumers (C₂)
200 000   primary consumers (C₁)

1 500 000   primary producers (P)

(a) **scale** ├──┤ = 100 000 individuals per 0·1ha

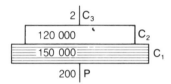

2 │ C₃

120 000   C₂

150 000   C₁

200 │ P

(b) **scale** ├──┤ = 20 000 individuals per 0·1ha

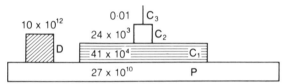

10×10^{12} $0·01$ │ C₃

24×10^3 │ C₂

D 41×10^4 C₁

27×10^{10} P

(c) Not to scale. Figures are numbers of individuals under $1m^2$ of water surface.

Pyramids of numbers for
(a) Derelict grassland in Michigan. (E. P. Odum, 1971.)
(b Temperate oak forest at Wytham Woods, England. (G. C. Varley (1970).
Symposium of the British Ecological Society **10**, 389–405.)
(c) Temperate river at Silver Springs, Florida. (H. T. Odum, 1957.)

FIGURE 70

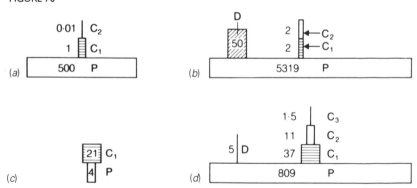

0·01 │ C₂

1   C₁

(a)   500   P

D

50   2 ◄ C₂
      2 ◄ C₁

(b)   5319   P

21 │ C₁

4 │ P

(c)

1·5 │ C₃

11   C₂

5 │ D   37   C₁

(d)   809   P

Pyramids of biomass in g m^{-2}. These drawings are not to scale.
(a) Derelict grassland in Georgia. (E. P. Odum, 1971.)
(b) Oak forest at Wytham Woods, England. (Varley, 1970.)
(c) English Channel. (H. W. Harvey (1950). *Journal of the Marine Biology Society
U.K. n.s.*, **29**, 97–137.)
(d) Silver Springs, Florida, a river. (H. T. Odum, 1957.)

sumers over the year. This in turn would exceed the biomass accumulated by the fish.

We could therefore draw pyramids of the total biomass accumulated by the organisms in a trophic level over the whole year. They would usually be pyramidal in shape. Yet two organisms with the same mass do not necessarily have the same energy content, for two reasons. Firstly, whereas carbohydrates and proteins contain $17\,kJ\,g^{-1}$ of potential chemical energy, fats possess $37\,kJ\,g^{-1}$. Secondly, some organisms contain a high proportion of inorganic salts. For example, 42% of the mass of the freshwater shrimp *Gammarus* consists of calcium carbonate in the exoskeleton. This does not represent energy trapped from the environment. For these two reasons it seems more satisfactory to draw pyramids of the energy entering each trophic level in an ecosystem over the whole year, particularly since energy flow is such an important feature of ecosystems. The figure which is plotted for the primary producers is the energy taken in by photosynthesis over the whole year. The figure for each of the consumer levels is the energy entering the mouths of the organisms each year.

The pyramids (figure 71) show that food chains in the tropics tend to be long, with five or six successive consumer levels. In arctic tundra only two or three consumer levels exist. Presumably the short growing season, and low intensity of solar radiation in the arctic, limit primary production and hence the energy available for consumers. Higher consumers, if they

FIGURE 71

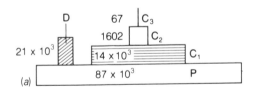

(a)

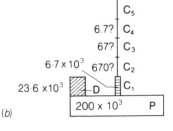

(b)

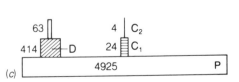

(c)

Pyramids of energy flow for three selected ecosystems. Units are $kJ\,m^{-2}\,year^{-1}$. Rectangles are not drawn to scale.

(a) Silver Springs, Florida – a large flowing stream.

(b) Tropical rain forest, El Verde, Puerto Rico.

(c) Arctic tundra, Devon Island, Canada.

For sources see page 174.

existed in the arctic, would tend, like other carnivores, to be smaller in numbers and larger in size than their prey. Their population density might be too low for individuals to find mates. Also, the energy expended in finding their sparse prey (snowy owl, arctic fox) might exceed the energy gained by assimilating it!

The pyramids of energy also show clearly that energy transfer between plants and their herbivores is much less efficient than between any other pair of trophic levels.

Rather more instructive than pyramids of energy are the energy budgets which can be drawn up for ecosystems.

ENERGY BUDGETS FOR ECOSYSTEMS

When the energy budgets for ecosystems are drawn in the same format they can be readily compared (figure 72). Study these diagrams with care.

There are three main types of ecosystem.

(a) Grazing ecosystems. In grasslands, fresh water and the sea, much of the net primary production is eaten by herbivores. The rest enters the detritus food chain when the plants die. Over the whole year the ecosystem is in balance. The energy entering it equals the energy leaving it.

In the temperate grassland in Sweden only 5% of the net primary production is eaten by herbivores. This proportion is usual when large mammalian grazers are absent. In units where cattle or sheep are reared in intensive milk or meat production the animals usually collect more than 5% of the P of the grassland. For example, the bullock illustrated in figure 65, p. 100 has eaten 3053/21 135 (14.4%) of the net primary production. This underestimates the energy entering the grazing food web because there are probably also many invertebrate herbivores in the ecosystem.

Silver Springs (figure 67, p. 103) is a good example of a grazing ecosystem.

(b) Detritus ecosystems. In mature heathlands and climax tropical and temperate forests a high proportion of the net primary production enters the detritus food chain. The grazing food chain is weak. Again, over the whole year, the energy captured by green plants in photosynthesis equals the energy lost in the 'respiration' of all the organisms in the community.

In the El Verde tropical forest (figure 72) a surprisingly high propor-

FIGURE 72 (right)
Energy budgets for eight selected ecosystems. Key at top of page. Losses of energy through 'respiration' are not shown but can be calculated from these data in many cases. Most of the numbers beside the arrows are percentages of net primary production. For sources see page 175.

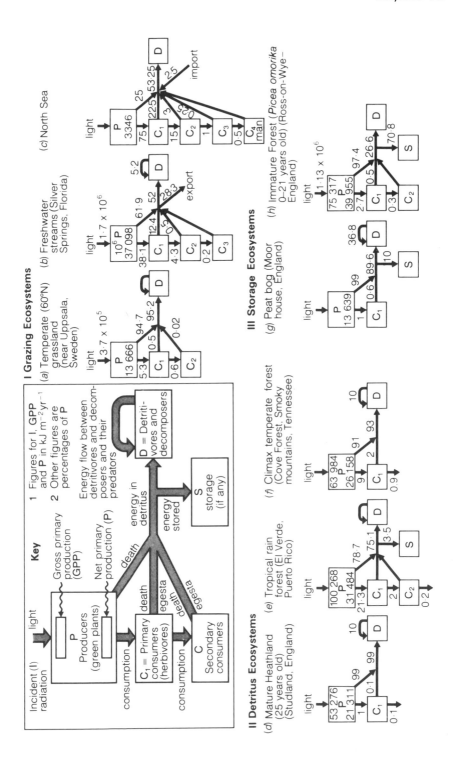

tion of the NPP enters the grazing food chain. Data on tropical rainforests are scanty and this may be a freak result.

(c) Storage ecosystems. In peat bogs, and heathlands and forests in an early successional stage, the energy entering the system each year exceeds the energy leaving it. The balance of the energy is stored. Of the net primary production, little is eaten because the grazing food chain is weak. Much of the rest is stored, either in peat in peat bogs or in the trunks, roots and branches of shrubs or trees in heathland or woodland.

In peat bogs the decomposers cannot readily decompose organic matter under acid anaerobic conditions. Much of the net primary production accumulates as peat. Peat 9 m deep has built up since the last glaciation in many raised bogs in the British Isles. When the peat is burnt the energy which has been stored is released as heat.

In scrub, the rate of photosynthesis is much greater than the rate of respiration and so the net primary production is high. Little of this is eaten. Much of the energy is deposited in trunks, branches and woody roots, often as dead lignified cells which are equivalent to peat.

As succession proceeds, and the scrub slowly turns into climax forest, its rate of photosynthesis hardly increases, because even the original scrub was intercepting a high proportion of the light energy. Yet this photosynthesis has to support an increasing volume of respiring tissue in the growing roots, branches, trunks and twigs. The net primary production therefore declines and new wood is laid down more slowly. Eventually, in a climax forest, the rate at which the trees store energy in new wood is balanced by the rate at which dead wood is broken down by detritivores and decomposers. The ecosystem as a whole stores no energy. A stable equilibrium is attained, as shown by the energy budget for climax forest (figure 72f).

These energy budgets complete our survey of energy flow in ecosystems. The other major activity in which ecosystems are involved is nutrient cycling.

NUTRIENT CYCLING

Whereas energy from the sun flows through ecosystems, and is lost to outer space as heat, nutrient elements cycle (chapter 1). Nutrients are absorbed by organisms from the soil or atmosphere, circulate through the trophic levels, and are released again to the environment. The levels of these essential elements in the soil can affect the productivity, distribution and abundance of plant and animal species. Nutrient cycles, and the way in which they are affected by humans, are therefore of considerable interest.

There are two main types of nutrient cycle. Firstly, in the global cycles, like those of water, carbon, and nitrogen, there is a large reservoir of the nutrient in the atmosphere. The gases released from one ecosystem may well be used by another. Secondly, in local cycles, like those of phosphorus and potassium (see p. 12), the atmosphere is not involved. The elements circulate between organisms and the soil or water.

Water, carbon and nitrogen are particularly important nutrients and their cycles will be dealt with in turn.

Hydrological cycle

The cycle of water in the biosphere is illustrated in figure 73, which includes estimates of the rates at which water is transferred from one place to another.

On average, each water molecule circulates every ten to fifteen days. This rapid cycle is driven by the sun. Solar radiation evaporates water from water surfaces, the soil and organisms. Solar energy drives the winds. The diagram shows that many ecosystems on land are partly dependent for their water supply on water evaporated from the sea, which is blown over land in clouds. It does not bring out the fact that the distribution of rainfall over the land surface is very uneven.

On the scale of the whole biosphere, there is so much water that the rates at which water is split by photosynthesis and formed again by cellular respiration can be ignored.

FIGURE 73

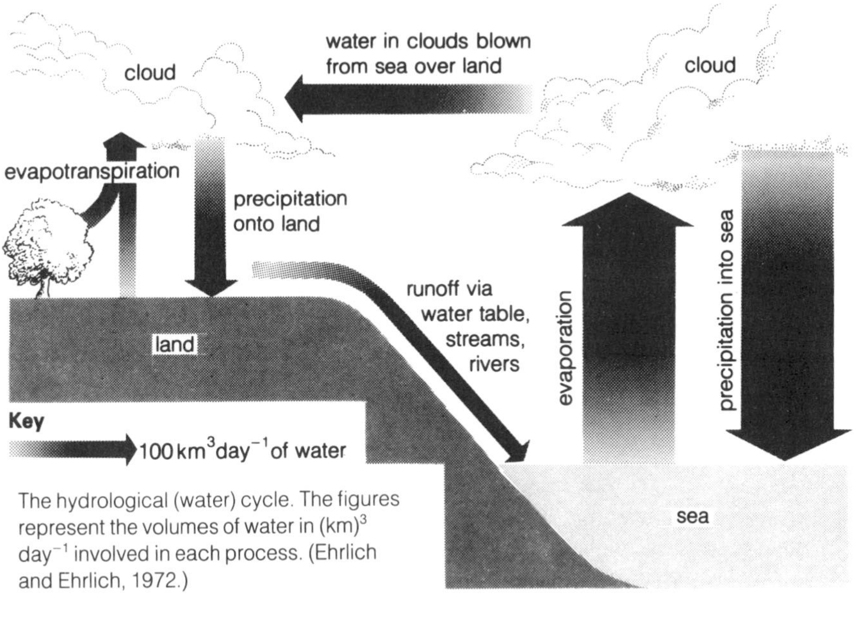

cloud

water in clouds blown from sea over land

cloud

evapotranspiration

precipitation onto land

runoff via water table, streams, rivers

evaporation

precipitation into sea

land

Key

$100 \, km^3 day^{-1}$ of water

The hydrological (water) cycle. The figures represent the volumes of water in $(km)^3$ day^{-1} involved in each process. (Ehrlich and Ehrlich, 1972.)

sea

The carbon cycle

In the carbon cycle (figure 74) there is a large reservoir of carbon dioxide in the atmosphere. This is exchanged both with the oceans, and with living organisms. Atmospheric carbon dioxide is in a dynamic equilibrium with the massive quantity of carbon dioxide stored in the sea as hydrogencarbonate ions. Carbon dioxide is also continually removed from the air by photosynthesis, and replaced by respiration.

Despite these processes, the level of carbon dioxide in the atmosphere is still rising (figure 75). This is because of the oxidation of stored organic compounds. The burning of fossil fuels creates carbon dioxide. So does the burning and decomposition of tree trunks and humus which follows the continual removal of large areas of forest.

In about 1850 the atmosphere contained about 270 ppm of carbon dioxide, according to analyses of old bubbles of air trapped in ice cores in the Antarctic. By 1957 its concentration had reached 315 ppm and now (1989) it is about 350 ppm and increasing at about 1 ppm a year.

The worst possible long-term effect of this increase in the atmospheric carbon dioxide concentration might be an increase in the average temperature of the earth's surface. This is because of the 'greenhouse effect'. The carbon dioxide in the atmosphere does not absorb the sun's short wave radiation, which then reaches the earth's surface. This energy is emitted by the earth as heat, some of which is absorbed by carbon dioxide and radiated back to earth. A blanket of carbon dioxide therefore keeps the Earth warm. The simplest hypothesis is that the more carbon dioxide in the atmosphere, the warmer the Earth should become.

Apart from carbon dioxide, the concentrations of methane (CH_4) chlorofluorocarbons (CFCs, p. 133) and nitrous oxide (N_2O) in the atmosphere are increasing and all these gases can contribute to the 'greenhouse effect'. For example, one molecule of CFC has the same warming effect as 10^4 carbon dioxide molecules. The methane concentration in the atmosphere is rising by nearly 2% each year because it is produced by the increasing populations of cattle and termites, and a greater area of paddy fields.

It is difficult to detect clear signs of global warming in the presence of natural fluctuations in climate. World average surface air temperatures increased by about 0.25°C between 1880 and 1940, but then decreased from

FIGURE 74 (right)
The cycle of carbon in the biosphere. (Figures from C. F. Baes Jr. *et al.* (1977). Carbon Dioxide and Climate: The Uncontrolled Experiment. *American Science*, **65**, 310–20, and G. M. Woodwell, 1978).

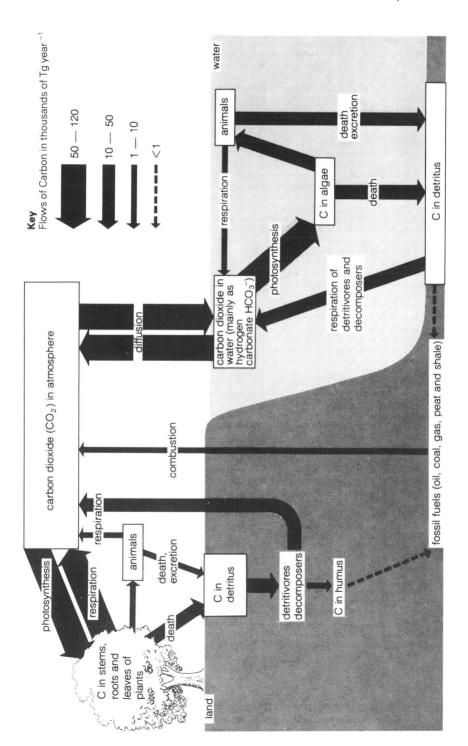

Key
Flows of Carbon in thousands of Tg year^{-1}

50 — 120
10 — 50
1 — 10
<1

water

animals

death
excretion

respiration

photosynthesis

C in algae

death

C in detritus

carbon dioxide in water (mainly as hydrogen carbonate HCO$_3^-$)

respiration of detritivores and decomposers

diffusion

carbon dioxide (CO$_2$) in atmosphere

combustion

fossil fuels (oil, coal, gas, peat and shale)

photosynthesis

respiration

respiration

animals

death, excretion

C in detritus

death

detritivores decomposers

C in humus

C in stems, roots and leaves of plants

land

FIGURE 75

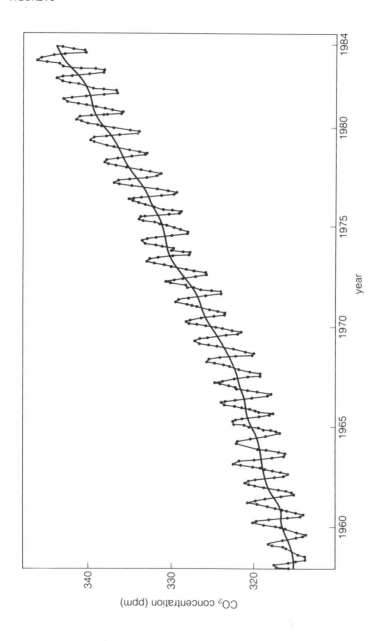

The concentration of carbon dioxide in the atmosphere, measured at the observatory at Mauna Loa, Hawaii, from 1957 to 1984. Notice the fluctuation in carbon dioxide level during each year. What causes this fluctuation?

1940 to 1970 by 0.2°C. Between 1970 and 1980 world temperatures increased by 0.3°C, and 1987 and 1988 were the warmest years on record (figure 76).

Yet the 'greenhouse effect' is a considerable oversimplification. Other important considerations are as follows.

(a) As the carbon dioxide concentration in the atmosphere increases, some of the extra carbon dioxide may be dissolved in the oceans. Alternatively, some may be stored in the growing forests, storage ecosystems (p. 109) which are replacing tropical forests or planted in the developed countries on land 'set aside' from agriculture.

(b) As the atmosphere warmed up, much of the heat might be transferred from the atmosphere to the oceans. The vast volume of water in the oceans would act as a massive heat sink, moderating the increase in temperature that might otherwise occur.

(c) An increase in the temperature of the earth would increase the rate of evaporation of water and hence the area of cloud cover. The extra clouds would reflect more solar radiation to outer space, compensating for the warming effect of the carbon dioxide.

These three factors would all tend to prevent an increase in the temperature of the atmosphere. On the other hand, it is just as likely that a slight

FIGURE 76

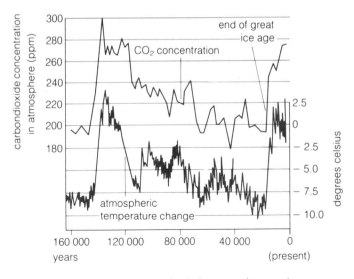

Correlation between carbon dioxide concentration in the atmosphere, and atmospheric temperature change, over the last 160 000 years. These data suggest that world climate and the carbon cycle are interrelated. The measurements were made on an ice core 2083 m deep from Vostok (East Antarctica). (J. M. Barnola et al. (1987). Nature **329**, 408–14.)

rise in the temperature might trigger off changes which increase the earth's temperature still further. For example, might a rise in the earth's temperature cause carbon dioxide to be released more rapidly by bacterial respiration in swamps and bogs? Once the contraction of the polar ice caps had begun, would the smaller area of snow and ice reflect less solar radiation, so that the earth warmed up at an even faster pace?

Most of the computer models (1989) predict a doubling in the effect of the greenhouse gases by about the year 2040. Sea levels have already risen by 15 cm since 1900, but a further increase in temperature would increase sea levels even more, as sea water expands and glaciers melt more rapidly. Low-lying areas would flood. Temperate regions would exhibit higher temperature increases than the tropics, and this, together with a reduction of rainfall in the centres of continents, might alter considerably the geographical distributions of the species and races of crop plants grown throughout the world.

The nitrogen cycle

Since nitrogen gas makes up 79% of the atmosphere, its atmospheric reservoir is over 2500 times larger than that of carbon dioxide. The nitrogen cycle is important because in many ecosystems a shortage of nitrogenous compounds reduces primary production and hence secondary production.

In the central cycle, nitrogen circulates between organisms and the soil (figure 77). Plants absorb nitrates or ammonium ions. They combine the nitrogen with carbon skeletons to form organic compounds like amino acids and proteins. These compounds may or may not be eaten by animals, but they eventually reach the detritus. There they were attacked by detritivores and decomposers, and the nitrogen is released as ammonium ions. Many of these ammonium ions are then converted by soil bacteria into nitrate, a process called 'nitrification'. The bacteria which carry out the conversion of ammonium to toxic nitrite, and nitrite to nitrate, are chemoautotrophic. In other words, they derive their energy from the oxidation of these inorganic ions. The central cycle is completed when the ions are taken up once more by plant roots.

There are three main ways in which ecosystems gain nitrogen from the atmosphere.

(a) Electrical discharges in the atmosphere combine nitrogen and oxygen. The input of nitrate ions from this source amounts to about 7 Tg year^{-1} over the whole of the earth's surface (figure 78).

FIGURE 77 (right)
The cycle of nitrogen in the biosphere.

The cycle of nitrogen in the biosphere.

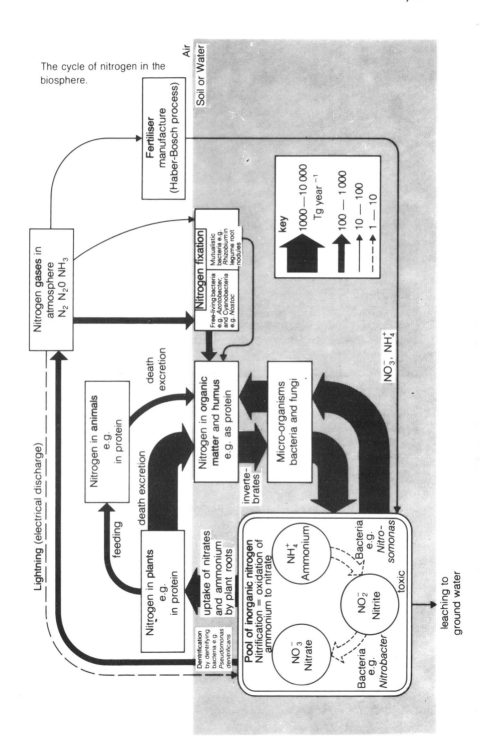

FIGURE 78

Soil and water gains from atmospheric N_2/N_2O

	$Tg\,year^{-1}$
Free-living N-fixing bacteria	169–269
Symbiotic N-fixing bacteria (legumes)	35
Fertiliser production (Haber–Bosch process)	76
Lightning	7
Total gains =	287–387

Atmospheric gains as N_2/N_2O from fixed N in soil and water

	$Tg\,year^{-1}$
Denitrification by denitrifying bacteria in	
waters:	91
land:	132–340
Total losses =	223–431

Balance sheet for nitrogen in the biosphere. All estimates are approximate. (Based on R. Soderlund and B. H. Svensson (1976). The global nitrogen cycle. *Ecology Bulletin* (Stockholm) **22**, 23–73; and C. C. Delwiche (1970). The nitrogen cycle. *Scientific American* **223**, 137–46.)

(b) Some bacteria and cyanobacteria can extract nitrogen gas from the atmosphere and convert it into inorganic nitrogen compounds. This process, called 'nitrogen fixation', requires energy from respiration or the sun. There are many free-living nitrogen-fixing micro-organisms. Others occur in mutualistic associations, such as the *Rhizobium* bacteria which occur in the root nodules of leguminous species like clover.

(c) The application of nitrogenous fertilisers like ammonium nitrate. Many of these fertilisers have been made by the Haber–Bosch process, in which atmospheric nitrogen gas is combined with hydrogen gas to produce ammonia. This industrial process contributes about a third of the total nitrogen fixation each year (figure 78).

The gain in nitrogen from these three sources seems to be balanced by denitrification, in which bacteria take up inorganic nitrogen compounds and release nitrogen gas to the atmosphere.

The nitrogen-containing fertilizers used in agriculture contribute to a pollution problem known as 'eutrophication'. The nitrates are washed out ('leached') by rainfall from agricultural fields into groundwater and streams. In some rivers in Britain the nitrate concentration in summer is fifty times that in 1950. Rivers and lakes have become enriched with nitrates and phosphates on a large scale in the developed nations, particularly in lakes bounded by cities and grain crops.

As a result of eutrophication the plants at the water's edge grow rapidly. Expensive clearance of water weeds may be necessary to allow water traffic to pass freely. In summer, the algae in the water undergo a population explosion called an 'algal bloom'. Rivers resemble pea soup instead of clear mountain streams. The algae have to be removed by water authorities before the drinking water reaches our taps. This requires a sophisticated and expensive filtration system.

When the algae in the river or lake die, they are decomposed by bacteria. Since the algae die in vast numbers, the bacteria undergo a population explosion. They take up oxygen in cellular respiration, deoxygenating the water and producing some toxic hydrogen sulphide gas. Those fish and invertebrates which require high levels of oxygen may die. In this way, eutrophication has already reduced the range of animal and plant species which can be found in many lakes and rivers.

High levels of nitrate in drinking water seem to be toxic to babies. They readily convert it in their blood or guts to toxic nitrite. In the great drought in Britain in the summer of 1976, many mothers were provided with nitrate-free drinking water to make up feeds for their babies, since pollutants in tap water had become highly concentrated.

NUTRIENT CYCLES AND FOREST DESTRUCTION

Experiments and observations on energy flow and nutrient cycling have greatly improved our understanding of most ecosystems. The effects of exploitation by humans can now be predicted more accurately. This is true in particular of the destruction of tropical rainforests (see also p. 148) and the removal of temperate forests.

The destruction of tropical rainforests

After the vegetation of a tropical rainforest has been removed the soil has a low nutrient content and will only support crops for two or three years. Then it becomes infertile. This is what happens in 'slash and burn' agriculture. The reasons for this is that most of the nutrient elements in the ecosystem are in the vegetation and only a small proportion are in the soil. When the vegetation is removed, the nutrient elements are lost from the ecosystem.

In a tropical rainforest about 80% of the organic carbon and 58% of the total nitrogen is in the biomass. The equivalent figures for a Scottish pine forest are 40% and 6%, respectively. In the constant high temperatures and moist conditions in tropical rainforests, the decomposition of dead leaves is very rapid and the uptake of nutrient ions by roots and mycorrhizal fungi is also very fast. Since the ions which leave the trees for the soil are rapidly re-absorbed by the vegetation, the nutrient content of the soil remains low (figure 79). In temperate and boreal regions, which usually have a lengthy cold season, the release of ions to the soil by decomposers and the uptake of nutrient ions by roots are relatively slow. The soil has a high nutrient ion content.

FIGURE 79

Area of zone $(m^2 \times 10^{12})$	Zone	Total nitrogen per unit area of litter $(g\ nitrogen\ m^{-2})$	Litter turnover time (half-life of litter nitrogen) (year)
8.1	Polar	106	66
23.2	Boreal	76	12
22.5	Temperate	11	1
24.3	Subtropical	12	0.4
55.4	Tropical	6	0.2
	Average for biosphere	26	1.3

Nitrogen in litter in different climatic zones. (Based on values given by N. I. Basilevich (1974). *Proceedings of the 1st International Congress on Ecology* (Wageningen: Pudoc) 182–6.)

Even if the vegetation is burnt on the site, its nutrients may rapidly be lost from the ecosystem. Some will blow away in smoke. Those ions left on the soil surface will be leached through the soil by rainfall or washed away, on a slope, by sheet erosion. Many of the remaining nutrients will be taken up by the crop. When the crop is harvested, some of these nutrients will never be returned to the ecosystem.

It therefore seems unlikely that tropical rainforest can ever be converted to productive agricultural land. Even the removal of the hardwood trees will drastically reduce future plant growth on the same site, since the nutrients in the harvested trees may take thousands of years to replace.

Any human interference in the fragile tropical rainforest ecosystem seems futile. The forests are 'deserts covered with trees'. The Brazilian government and private 'developers' of the Amazonian rainforest have learnt this lesson the hard way. Their original plan, against the advice of ecologists, was to exploit the 2.7 million km^2 of forests for agricultural production and forest products. To this end they drove 22 000 km of roads through the region and set up agricultural settlements on cleared sites. The roads and settlements merely created vast tracts of sterile land. The emphasis has now shifted to ecological research on untouched forest, which might suggest ways in which the forest can be carefully exploited without affecting its long-term integrity and productivity.

Temperate forests

Considerable losses of nutrients also occur when temperate forests are felled. Much of the research has been carried out in the Hubbard Brook system of New Hampshire, which is a group of deciduous forest watersheds dissected by streams. Since these forests are on impermeable granite, nutrient losses to the bedrock are probably negligible. All the nutrients from the ecosystems reach the outflow streams, where the concentrations of ions can be measured (figure 80).

In 1966 the trees at Watershed Two, covering 15.6 ha, were cut down and left on the soil surface. Although the soil was disturbed as little as possible, the effects were striking. The volume of water running off the site over the next three years was 26–41% greater than the water loss from an unfelled control watershed nearby. The streams leaving the felled forest contained much higher concentrations of ions than the stream from the unfelled forest. Nitrate levels were 40–60 times higher, potassium 15 times, calcium and magnesium both four times, and sodium 1.8 times.

In an intact forest much of the rainfall is evaporated from the foliage and branches of the trees, without reaching the soil. After a forest has been felled, more water reaches the soil surface. This increased volume of water leaches more of the nutrient ions into streams, and on steep slopes, may

FIGURE 80

Watershed Two in the Hubbard Brook Experimental Forest, after felling. The gauge-house and weir are obvious in the foreground. (Photograph by Gene E. Likens.)

wash off the topsoil. The higher temperatures near the soil surface increase the rate of bacterial activity. This causes faster breakdown of humus and faster nitrification. The extra nitrate produced is leached away.

Thus the felling of a forest, even in temperate regions, may reduce the fertility of the soil.

It is clear from this chapter that humans are increasingly disrupting the natural patterns of energy flow and nutrient cycling. This is not surprising, since the human population is growing so fast (chapter 7). The by-products of industry also have major effects on ecosystems (chapter 8). The attempts to maintain some areas of semi-natural wilderness, and the arguments for doing so, are dealt with in the final chapter on Nature Conservation (chapter 9).

7 *The expanding human population*

There are over five thousand million humans on Earth. The population is increasing by about 150 each minute, and over 200 000 a day. Why are our numbers expanding so rapidly, and what, if anything, can be done about it?

At present most of the rapid population increase takes place in the less developed countries of the 'Third World'. Population growth in the developed countries has almost stabilised (figure 81). One way to look at this phenomenon is to think of human population increase as a three-stage process, known as the 'demographic transition'. The less developed countries are still in the second phase, but the developed nations have reached the third. In *stage I*, both birth rates and death rates are high, and the population only increases slowly. In *stage II*, birth rates remain high, but because of improvements in sanitation, nutrition and medical care the death rate falls. There is a rapid population increase. In *stage III*, the birth rate falls because of family planning. Ultimately the population size stabilises. Both birth rates and death rates are low. The various stages in this process are apparent in figure 82 for both developed and less developed countries.

THE HISTORY OF HUMAN POPULATION GROWTH

The earliest skulls and bones of human-like animals, found in the Rift Valley of Kenya, are two or three million years old. The human fossil record is scanty, but the earliest skulls and teeth which closely resemble those of modern man (*Homo sapiens*) are those of Cro-Magnon man, found in the Dordogne in France. Radiocarbon dating (see p. 81) suggests that the remains are about 32 000 years old. These people made tools and weapons, produced the first known art and sculpture, created many exquisite cave paintings, and were probably hunters of animals and gatherers of wild fruits. They and their descendants may have hunted to extinction mammals like woolly mammoths and sabre-toothed tigers.

The changes in human population size over the last 32 000 years can be crudely estimated from the number and geographical distribution of archaeological sites which contain the weapons and tools of each period.

FIGURE 81

	Population 1981	Population 2020 (est.) (millions)	Births per 1000	Deaths per 1000	Infant deaths per 1000	Total births per woman (est.)	% aged 15 years	% aged 65 years	Income per head (£year⁻¹)
West Germany	62	49	10	12	13	1.5	20	15	13590
Sweden	8.3	7.4	12	11	7	1.7	20	16	13520
U.S.A.	232	274	16	9	12	1.9	23	11	11360
U.K.	56	56.5	14	12	12	1.9	22	15	7920
U.S.S.R.	270	346	18	10	36	2.3	24	10	4550
Mexico	72	140	32	6	56	4.8	42	3	2130
Ecuador	8.5	23	42	10	82	6.3	45	4	1220
Kenya	18	59	53	14	87	8.1	50	4	420
China	1000	1400	22	7	45	2.8	32	6	290
Gambia	0.6	1.8	49	21	198	6.4	42	2	250
India	714	1197	35	15	123	5.3	40	3	240
Burma	37	76	39	14	101	5.5	40	4	180

Population characteristics of selected countries in 1981. The countries are arranged in order of decreasing income per person. Notice the trends in birth rates, infant mortality, family size and the proportions of youngsters and oldsters. (World Population Data Sheet (1982) of the Population Reference Bureau Inc., Washington D.C.)

FIGURE 82

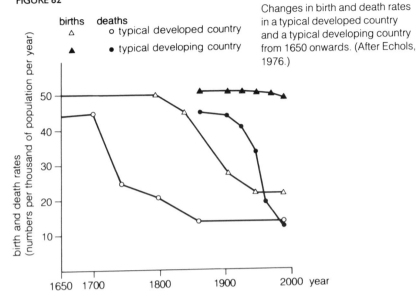

births deaths
△   o typical developed country
▲   ● typical developing country

Changes in birth and death rates in a typical developed country and a typical developing country from 1650 onwards. (After Echols, 1976.)

The world population rose from five million, ten thousand years ago to two or three hundred million at the time of Christ. By 1650 there were about five hundred million humans. Now there are over five thousand million. There seem to have been two major spurts in population growth, one about ten thousand years ago and the other beginning about 1650.

The first rapid increase was sparked off by the development of agriculture, about 9000–11 000 years ago, in the 'fertile crescent' of the Middle East, between Iraq and Iran. Goats, sheep and cattle were domesticated and bred. The 'seeds' of wild grasses were sown on bared ground and the grains — strictly fruits, not seeds — were collected. Grain produced in favourable seasons of the year could be stored, and eaten when food was scarce. Crop farming involved continuous cultivation. Nomadic tribes began to settle, and villages and towns sprang up. Agriculture could support a larger population than hunting and gathering, and the population increased.

The second rapid spurt in numbers began about three hundred years ago. The increase in the world's population since 1650 is shown most dramatically when the number of humans is plotted on a linear scale against time (figure 83). There were about five hundred million individuals in 1650, but by 1927 the population had quadrupled. By 1975 it had doubled again to reach four thousand million.

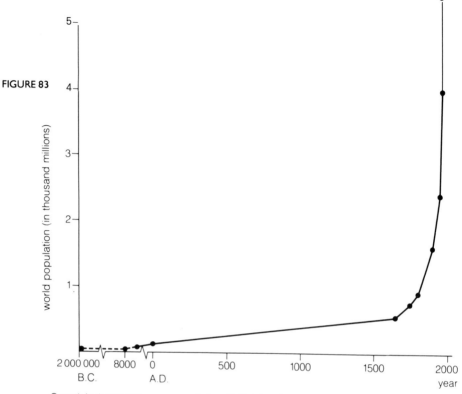

FIGURE 83

Growth in the world human population. (Initial data from E. S. Deevey Jr. (1960). The Human Population. *Scientific American* **203**, 194–204.)

THE CAUSES OF THE POPULATION EXPLOSION

Since 1700 the expectation of life has increased considerably, particularly in the developed countries (figure 84). In particular, the chance that a child will die in infancy has been very much reduced. Because of this lengthened life expectancy, more of the population now reach reproductive age and have children. This is the main cause of the recent substantial increase in human numbers.

FIGURE 84

Approximate mean life expectancies of humans in Europe at different periods in history. I. M. Lerner (1968). *Heredity, Evolution and Society*. (Freeman, San Francisco.)

Period	Human life expectancy at birth (years)
8000–3000 B.C.	18
A.D. 800–1200	31
A.D. 1600–1700	33.5
A.D. 1800–1900	37
Early 1900s	57.4
Mid 20th century	66.5

The expectation of life of an individual has increased because of the reduction of disease and the increase in the food supply. Let us look at each in turn.

The role of viral and bacterial diseases in causing mortality is illustrated by the causes of death amongst the inhabitants of London in 1641 (figure 85). A high proportion of the deaths were the result of ague and fever, consumption (bacterial tuberculosis of the lungs), smallpox, the plague, spotted fever, and 'teeth and wormes'. The plague (the 'Black Death'), killed a quarter of the inhabitants of Europe between 1348 and 1350. From 1348 to 1379 the population of England was reduced from 3.8 million to 2.1 million.

Several important scientific advances have reduced disease. Firstly, immunisation and vaccination on a large scale have dealt a considerable blow to infections such as smallpox, yellow fever, tuberculosis and poliomyelitis. Secondly, anti-bacterial drugs like penicillin can now be used to cure bacterial infections. Thirdly, and probably most importantly, there have been considerable improvements in sanitation and public health. Rubbish and sewage began to be disposed of in places where they did not contaminate drinking water. Piped water was supplied to towns. Dysentery, typhoid and gastroenteritis became less frequent. Higher standards of cleanliness have reduced the population sizes of many vectors of disease, such as fleas, lice and rats. Research into the natural history of many diseases has pinpointed the alternative host of many human parasites. Effective attacks have been made, for instance, on the snail which carries bilharzia (*Schistosoma* spp.) and the mosquitoes (*Anopheles* spp.) which carry malaria (*Plasmodium* spp.).

Apart from the control of disease, the increase in world food production has fuelled the population explosion. More land has been brought into cultivation. European settlers colonised the Americas and their populations expanded rapidly. They sent back goods and foodstuffs to their native countries. This raised the standard of living and the quality of nutrition in Europe. Agriculture became more efficient, thus increasing both the quality and the quantity of food. Farms became larger and were run as businesses. Crop strains of animals and plants were developed. The value of clover in enhancing soil fertility was recognised. Crop failures and famines became less frequent. The industrial revolution has provided machines which allow a single human to cultivate large areas at a time. Yields have been increased by the application of artificial fertilisers. Synthetic herbicides, insecticides, fungicides and molluscides can be used to kill pests which might otherwise reduce the mass or quality of the crop. Finally, our understanding of human nutrition has improved. Many of the symptoms of malnutrition can be relieved by supplementing the diet with protein, essential amino acids, or vitamins.

FIGURE 85

A general bill of mortality for London, for the year ending 16 December 1641. Notice the causes of death. (Guildhall Library, City of London.)

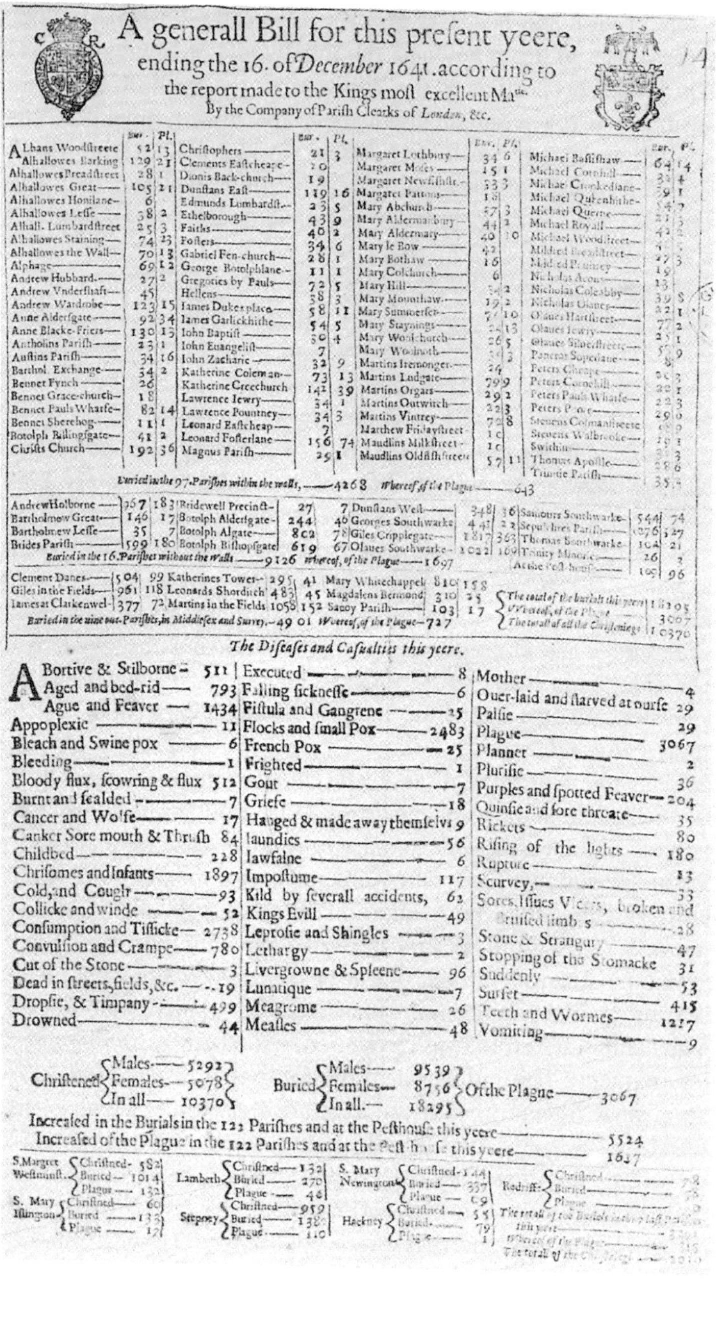

Thus over the last two hundred years there has been a transition from a population with a high infant mortality, in which the rate of increase was limited by disease and food shortage, to a population which exploded once these constraints had been reduced.

POPULATION STABILISATION

In the last fifty years, however, family size has decreased dramatically in most of the developed countries (figure 81). In Britain, for example, the average family had 7.6 children in 1890 but 2.3 children in 1975. Most of the developed nations now have fairly constant populations in which the birth rate has declined to the level of the death rate. The main cause of the recent decline of the birth rate is the widespread knowledge, acceptance, and use of methods of birth control. This enables couples to choose how many children to have.

Why have couples in the developed countries opted for smaller families than before? There are probably two main reasons. Firstly, infant mortality in the developed nations is now so low that parents can be sure that most of their offspring will grow to maturity. Previously, the survival of children was unpredictable and a large family size increased the chances that some children would live to adulthood. Secondly, children in industrial economies are consumers and cost a lot to feed and clothe. Previously, in rural economies, they were regarded as potential producers because they could work the land. Most families in the developed countries now have the freedom to choose whether to spend their money on children or on material goods.

In contrast, in many of the developing nations of the 'Third World', the decline in the death rate, and the resulting population explosion, have only occurred within the last fifty years. These countries have 'benefited' only recently from industrialisation and increases in agriculture and medicine. The rapid population growth puts a considerable economic strain on the nations concerned. A high proportion of the expanding population are young (figure 86) and do no work. Yet they must be fed. If money is used to provide better educational and medical facilities for the children, it is diverted from agricultural and industrial development, which could create wealth and provide jobs for the youngsters when they grow up.

Many family planning programmes have been started in the developing countries in an attempt to reduce family size. The difficulties are immense. There is frequently a shortage of doctors, trained medical staff, and equipment. Contraceptive techniques often have to be explained to people who are suspicious, illiterate and unsophisticated. Many people still wish to have large families. In some countries the government or the church is opposed to contraception.

As a result, family planning programmes have so far made a small

FIGURE 86

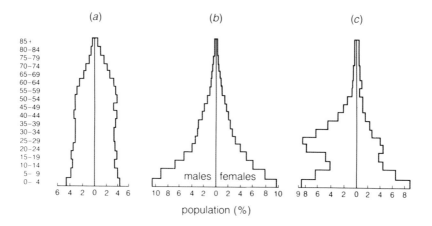

Three age pyramids for human populations: (a) United Kingdom (1965), (b) Costa Rica (1963), (c) Kuwait (1965). Notice the low birth and death rates for the United Kingdom population, the high proportion of children in Costa Rica and the effect of male immigration (largely to service the oil industry) on Kuwait's population. (M. K. Sands (1978). *Problems in Ecology*. Mills and Boon.)

impact on the worldwide population increase. Yet in the long run, family planning propaganda may create a climate which hastens the day when the birth rate in the developing countries falls to the same level as the death rate.

There have already been some successes. Both China and India have reduced their birth rates. China has banned marriage to women under twenty-five years old. India has provided bribes like transistor radios to men who are sterilised by vasectomy or women who are fitted with intra-uterine contraceptive devices. Nevertheless, it seems likely that even in these countries, the population explosion might continue for at least a century.

Imagine the unlikely case that India and China underwent such rapid industrialisation, and their family planning programmes worked so well, that by A.D. 2000 the average family had 2.5 children. Their populations would still increase, for two reasons. Firstly, the large numbers of children produced between 1980 and A.D. 2000 would enter the reproductive age groups between 1995 and A.D. 2015. This would ensure that the number of couples producing children might increase up to about A.D. 2025. Secondly, because of continuing improvements in medical care, these children would stand more chance than before of surviving to reproductive age and themselves producing children. Yet this is the most favourable possible case, and

India and China between them have nearly 40% of the world's population! In many developing countries where the population is increasing rapidly, industrialisation, literacy and family planning programmes, all of which seem to lower the birth rate, seem much further away than in India or China. In the poorest nations we might expect the populations to increase for at least two centuries.

There are two encouraging signs. The first is that where birth control programmes have been tried in the Third World on a large, well-organised scale, they have been markedly successful. The second is that in terms of the typical sigmoid (S-shaped, logistic) curve of population increase for a species invading a new habitat (figures 30 & 31, pp. 48 & 49), the human population is in the 'log phase' of growth but has passed the steepest point on the curve. The percentage increase is slowing down, year by year, in both developed and less developed countries (figure 87).

FIGURE 87

Year	World	Percentage increase in: More developed countries	Less developed countries
1955	1.84	1.28	2.11
1960	1.86	1.27	2.14
1965	1.96	1.19	2.30
1970	2.06	0.97	2.85
1975	2.03	0.89	2.46
1980	1.77	0.74	2.14
1985	1.67	0.64	2.02

Percentage increases in the world human population each year. Notice the progressive decline in population increase in the more developed countries. Population increases in the world, and the less developed countries are now also slowing down. (Data from United Nations Environmental Data report (1987). Blackwell.)

The world's population must ultimately level off or begin to decline, because the resources of the planet are finite. The classic constraints of war, famine, disease, or the accumulation of toxic wastes may ultimately limit human population size. War is a political force and difficult to predict. Yet there is certainly a limit to the sustained yield of crops which humans can grow, and the increasing population density of humans no doubt increases the chance of a pandemic. AIDS is a possibility. Many ecologists feel that in view of these inevitable constraints it might be best to stabilise the world population at as low a level as possible. This would reduce the inevitable suffering of millions of people. It would also allow much of the human population to live in a dignified and prosperous manner at a level which environmental resources could support in the long-term.

8 | *Pollution*

Pollution is strictly the release into the biosphere by humans of elements or compounds which have a harmful effect on humans or wildlife. This definition does not cover natural disasters, like the reduction in the earth's temperature for a year which was caused by the dust from the explosion of Krakatau (p. 40), or the natural leakage of oil from the earth's crust into the sea. Nor does it include 'noise pollution'. Nevertheless, it is sufficient for most purposes.

Pollutants can be classified into three main types.

(a) Some substances are effective pollutants soon after they are released, but rapidly disappear. The threat is removed by physical or chemical means. Carbon monoxide, for example, is converted to carbon dioxide by chemical reactions or by bacteria. The hot water which enters estuaries from power stations, and causes thermal pollution, soon cools down.

(b) Some pollutants such as sewage and oil are 'biodegradable' because they are broken down by living organisms.

(c) Many ions and compounds, such as cadmium and the organochlorines, cannot be broken down by organisms and are called 'non-biodegradable.' These substances are accumulating in the biosphere and probably represent the most important long-term threat to living organisms.

In order to illustrate the biological and economic effects of pollutants, sulphur dioxide and acid rain, organochlorine pesticides, radiation and chlorofluorocarbons will be discussed in turn. These and other main pollutants are listed in figure 88 in which their main sources and effects on organisms are summarised.

SULPHUR DIOXIDE AND ACID RAIN

The acidic gas sulphur dioxide is released when fossil fuels like coal and oil are burnt. It is emitted from power stations, domestic chimneys and car exhausts. It may be converted in the atmosphere to ammonium sulphate, an acidic compound, or it may react with water, forming hydrogen sulphite or

FIGURE 88
The major pollutants and their effects.

(a) Pollutants rapidly removed from the biosphere by physical or chemical means

Pollutant	Source	Effects
Carbon monoxide (CO)	Gas produced by car exhausts in incomplete combustion of petrol. City streets, traffic jams. Oxidised to CO_2 in upper atmosphere.	Affects humans. Combines with haemoglobin in blood, forming carboxyhaemoglobin, which does not dissociate. Deprives brain of O_2. Causes faintness, slow reactions and ultimately death.
Carbon particles	Car exhausts, burning of coal, oil.	Blackens buildings, contributes to lung diseases.
Carbon dioxide (CO_2) — see p. 112	Combustion of fossil fuels, fires, decomposition of organic matter, respiration.	May contribute to 'greenhouse effect' — see p. 112.
Methane (CH_4)	Bacteria in marshes, paddy fields, guts of termites and ruminants.	May contribute to 'greenhouse effect' — see p. 112.
Sulphur dioxide (SO_2) — see p. 132	Combustion of fossil fuels, car exhausts.	Forms 'acid rain', kills lichens, etches buildings, exacerbates asthma, bronchitis — see pp. 132–5.
Oxides of nitrogen (N_2O, NO, NO_2)	Car exhausts, aircraft, peat bogs, denitrifying bacteria.	Poisonous. May deplete ozone layer. Contribute to 'acid rain'.
Ozone (O_3)	Car exhausts, electrical installations.	Damages plant leaves.
Peroxyacyl nitrates (PAN)	Formed in high light intensities from exhaust gases (N oxides, O_3 etc.) of cars. Accumulates in temperature inversions.	Choking 'photochemical smog' in Tokyo, Los Angeles, Sydney.
Chlorofluorocarbons (CFCs)—see p. 140	Propellants in aerosol cans, plastic foam blowing, refrigerator coolants.	Deplete ozone layer (upper atmosphere), allowing damaging UV rays to reach organisms, contribute to 'greenhouse effect' — see p. 112.
Fluorine (F_2), hydrogen fluoride (HF)	Brickworks, aluminium smelters.	Fluorosis (bone disease) in cattle. Reduces plant growth; damages leaves.

FIGURE 88 (continued)

Pollutant	Source	Effects
Thermal pollution	Hot water from cooling ducts of power stations, reaching rivers, estuaries.	Increases metabolic rate of organisms and lowers O_2 concentration in water. Reduces number of species. Migrating fish cannot travel across barrier of warm water. Lower salinity of water on local scale.
Eutrophication (nitrates & phosphates) — see p. 118	Fertiliser run off, sewage, detergents.	Excess growth of waterway vegetation, algal blooms, deoxygenation of waterways by bacteria, fish death — see p. 119.

(b) Biodegradable pollutants

Sewage	Excreta of humans and farm animals.	Bacteria decompose organic matter in untreated sewage and deoxygenate water, killing some fish and O_2-requiring animals. Intestinal bacteria may cause disease. Phosphates in sewage & detergents contribute to eutrophication (see p. 118).
Oil	Holed tankers (e.g. Torrey Canyon 1967, Amoco Cadiz 1977). Tankers washing out tanks at sea. Oil terminals. Leakages from earth's crust.	Biodegradable, by bacteria. Oils and kills seabirds. Kill algae and intertidal animals. Economic damage to holiday resorts. Affected areas are recolonised by wildlife over 5–10 years.

(c) Non-biodegradable pollutants

Lead (Pb)	Car exhausts; added to petrol as lead tetra-ethyl (anti-knock) to improve octane rating. Lead piping to houses supplying acid water. Lead paint. Spoil heaps near old lead mines.	Cumulative poison in brain; enters bloodstream through lungs (particulate lead) or gut. Mental retardation and death at high concentrations in humans. Lead from spoil heaps kills wildlife in streams and rivers.
Mercury (Hg)	Plastics factories. Factories making Na and Cl_2 from brine by Castner–Knellner process. Wood pulp factories. Seed dressing. Converted to easily absorbed methyl mercury by bacteria on ocean floor.	Accumulates up food chains. Top predators particularly at risk. Cumulative brain poison, causing nervous discoordination, mental retardation and death (e.g. Minimata bay, Japan 1955). Prevents nitrification in some seas.

FIGURE 88 (*continued*)

Pollutant	Source	Effects
Cadmium (Cd)	Aluminium and zinc smelters. Battery factories.	Causes disappearance of cartilaginous part of bone. Aching joints, brittle bones (itai itai disease in Japan). Kidney damage. Accumulated by shellfish in estuaries.
Polychlorinated biphenyls (PCBs)	Plastics waste dumped at sea by ships.	Seabird kills (e.g. Irish sea 1969).
Radiation — see p. 140.	Disintegration of nuclei of radioactive atoms. Testing of A- and H-bombs in atmosphere produces deposit of radioactive dust containing ^{90}Sr, ^{121}Kr, ^{91}I over wide area ('nuclear fallout'). Waste from nuclear power stations. Cosmic rays. Ultra-violet light.	Causes genetic mutations; increases the chances amongst humans of leukaemia and other cancers; effect is proportional to the cumulative dose. Isotopes may be concentrated up food chains — see p. 140.
Dioxins	Plastics and herbicide manufacture.	Very poisonous, developmental abnormalities in children.
DDT & other organochlorine insecticides — see p. 137.	Insecticides on crops.	Accumulate up food webs to toxic levels — see p. 138.

FIGURE 89

Open-topped chambers in which the effects of various atmospheric pollutant gases on tree growth are tested. The pumps on the right force air, contaminated or not, into the bases of the chambers. (Courtesy J. Lee, University of Manchester.)

hydrogen sulphate. In these forms it etches stone buildings. It also affects lichens, crop plants and humans.

Lichen species differ in their tolerances to sulphur dioxide. The lichen *Lecanora conizaeoides*, for example, the commonest lichen on trees, wood and brick in polluted areas, can tolerate 150–160 µg m^{-3} of sulphur dioxide in the air during the winter. At the other extreme, the beard lichens, species of the genus *Usnea* which trail from the branches of trees, can only be found in unpolluted habitats. They will die if the winter levels of sulphur dioxide in the atmosphere exceed 35 µg m^{-3}. Since many lichen species are susceptible to high sulphur dioxide levels, it is not surprising that in large industrial cities like Manchester and Newcastle only one or two lichen species can be found in the centre. Up to 150 species occur in graveyards 15 km away.

In fact the range of lichen species which grows in an area provides the best indication of the average level of sulphur dioxide in the air. Surveys of the sulphur dioxide levels in the air throughout Britain, using lichens as pollution monitors, have shown that some parts, like Scotland and the west coast, are relatively unpolluted, whilst industrial cities and the countryside downwind of them suffer from SO$_2$ fallout.

Periods of maximum suffering among humans who suffer from asthma and bronchitis in cities are correlated with higher than average levels of sulphur dioxide in the air. The gas is at least partly to blame for the 4000 extra deaths which occurred in the London 'smog' of 5–9 December 1952. Visibility was reduced to five or ten metres, as effluents accumulated in air which could not rise because of a temperature inversion.

The smogs stimulated the introduction of the Clean Air Act (1956), which allowed local authorities in the British Isles to set up smokeless zones in cities. The results have been dramatic. In London the concentration of carbon particles at street level was six times lower in 1977 than in 1950. In London and Manchester several lichen species, besides the fungus disease 'black-spot' of roses, and the white morph of the peppered moth (*Biston betularia*), have all moved nearer the centre during the last thirty years.

Despite these encouraging signs, sulphur dioxide is still a major pollutant. Each year 5 Tg are released into the atmosphere in Britain when fossil fuels are burnt! Since the mid 1950s, factory chimneys have been built two or three times higher. This tends to spread the sulphur dioxide over the surrounding countryside instead of allowing it to accumulate near the chimney at ground level. This widespread sulphur dioxide makes a contribution to 'acid rain'.

Acid rain is a mixture of gases which are (*i*) deposited directly on soils or plants (sulphur dioxide, nitrogen oxides, ozone, ammonia) near the pollution source, a phenomenon known as 'dry deposition', (*ii*) dissolved in water, forming an acidic rainfall containing sulphuric and sulphurous acids, nitric acid and ammonium sulphate, known in contrast as 'wet deposition'. Uncontaminated rain might have a pH of 5.6 but some rainfall in Scotland has had a pH as low as 1.6!

There is now sound evidence from Scandinavia and Scotland that lakes in which the fish have declined or disappeared have shown a marked acidification which began after the industrial revolution. This comes from an examination of the changes in the diatom species with depth in cores of sediment taken from the bottom of the lakes. Different diatom species differ in their tolerances to the pH of the water.

Experiments in Norway suggest that acid rain may be the cause of lake acidification. At Risdalsheia, for example, two forest catchments have been roofed over. One is sprayed with the usual acidic rainfall and the other with the acidic rainfall from which many of the ions have been removed by an ion exchange column. The removal of the pollutants has reduced the nitrate, sulphate and aluminium outputs from the forested area. In fact the increased levels of soluble aluminium ions in the lakes may be the main cause of fish decline. Thankfully, other experiments indicate that the symptoms of lake acidification can be reversed by liming the water.

There is concern in many European countries that some forests possess trees which look sick. Species such as Norway spruce (*Picea abies*) and Silver fir (*Abies alba*) exhibit loss of needles or leaves, death of the upper twigs and a yellowing of the older foliage in a phenomenon known as 'forest decline'. Many different ideas have been suggested to explain this phenomenon, and most of them involve an interaction of acid rain with other factors, such as magnesium deficiency in the soil. However, no general conclusions have emerged.

Acid rain is an international problem. Sweden, for example, receives most of its acid deposition from other countries. Experiments in which radioactive sulphur dioxide has been released from chimneys in Britain, and 'caught' in samplers on aeroplanes flying between Britain and Scandinavia, suggest that Sweden receives about 10% of its sulphur dioxide deposition from the United Kingdom. The problem will partially be solved by the installation of devices to remove acidic gases from the effluent from power stations and car exhausts, but only at considerable cost, directly or indirectly, to all of us.

ORGANOCHLORINE INSECTICIDES

The man-made organochlorine insecticides like DDT were introduced in the mid-1940s. In 1945 DDT was remarkably effective in ridding American soldiers in Italy of the head louse (*Pedicularis capitata*), which transmits typhus (*Rickettsia prowazekii*). Since then the organochlorines, sprayed from aircraft, have been used successfully on a large scale to rid tropical swamps of malarial mosquitos and to kill insect pests in agricultural crops. Yet their widespread use in Britain and America has been banned since the mid-1960s as a result of deaths in natural animal populations and the accumulation of DDT in food chains.

From the late 1940s to the early 1960s organochlorines were used in sheep dips and sprayed onto crops at an increasing rate. It was soon noticed that animals at the top of food chains sometimes died unexpectedly in areas where these pesticides had been applied. One well documented example is that of Clear Lake, California. The lake was treated with a low concentration of DDD in 1949, 1954 and 1957 to rid it of the local midges. The effect on the grebes was dramatic. Although there were 100 pairs of grebes on the lake in 1950, none of them hatched out a chick between 1950 and 1962. Many grebes died after the treatments in 1954 and 1957. Many similar cases were brought to public attention, particularly by Rachel Carson in her book 'Silent Spring' (1962). At about this time several birds of prey, such as the osprey and peregrine falcon, declined in numbers drastically in Britain and America. The public outcry eventually led to a ban on organochlorines, and their replacement by organophosphorous compounds.

DDT becomes more concentrated in organisms with each successive link in a food web (figure 90). This effect is particularly marked in aquatic ecosystems. It occurs because DDT and its derivatives accumulate within body fat and are hardly excreted. At Clear Lake its concentration in plankton was 250 times its original concentration in the lake water. On the same scale, the body fat of frogs had 2000 times, sunfish 12 000 times and grebes 80 000 times the concentration of DDT in the lake. Presumably each sunfish had accumulated the DDT residues from thousands of planktonic algae, and each grebe had eaten several contaminated frogs and sunfish.

FIGURE 90

	DDT residues (ppm)	
Water	0.000 05	The concentrations of DDT in the animals of a Long Island estuary. (G. M. Woodwell, C. F. Wurster and P. A. Isaacson (1967). DDT residues in an East Coast estuary: a case of biological concentration of a persistent insecticide. *Science* **156**, 821–4.)
Plankton	0.04	
Mud snail (detritus feeder)	0.26	
Clam (filter feeder)	0.42	
Silverside minnow	0.23	
Sheepshead minnow	0.94	
Pickerel (predatory fish)	1.33	
Needlefish (predatory fish)	2.07	
Heron (feeds on small fish)	3.57	
Tern (feeds on small fish)	3.91	
Herring gull (scavenger)	6.00	
Osprey (fish-eating bird) egg	13.8	
Merganser (fish-eating duck)	22.8	
Cormorant (feeds on larger fish)	26.4	

The levels of DDT in the body fat of top predators may be high enough to kill them or to prevent them from breeding.

The decline of the peregrine falcon (figure 91) in Britain during the 1950s was caused by a marked lack of breeding success. The eggs broke when the female birds sat on them. Three pieces of evidence suggest that organochlorine pesticides were responsible. Firstly, the concentration of DDT in the body fat of dead peregrine falcons was high. Secondly, from 1948 onwards the thickness of peregrine eggshells decreased, although measurements on museum specimens had shown that eggshell thickness remained constant from 1900 to 1948. Thirdly, experiments on female pigeons show that DDT affects the oestrogen levels and this in turn reduces the number of calcium ions deposited in the eggshell.

Organophosphorous insecticides like Malathion and Parathion, and carbamates such as Carbaryl, break down more rapidly than DDT. They are now used instead of the organochlorines on a large scale to kill crop pests. Nevertheless, the whole biosphere is still contaminated with DDT residues. DDT has been found in the body fat of penguins in Antarctica,

FIGURE 91

Peregrine falcons. This species declined rapidly in numbers in the British Isles after the introduction of organochlorine pesticides.

although it has never been used within the region. The concentration of organochlorines in the sea should slowly decline, but they will be a normal constituent of organisms for some time.

RADIATION

Various sorts of short-wave electromagnetic radiations (γ-rays, β-particles, α-particles) are released by the decay of radioactive elements. That they are capable of causing cancers and leukaemias is shown by the long-term studies on the Japanese who were unlucky enough to be near the blasts from the atomic bombs dropped on Hiroshima and Nagasaki in 1945. The nearer to the centre of the blast they were, the greater the probability that the survivors later contracted one of these diseases.

The testing of nuclear weapons in the atmosphere, and the release of radioactive isotopes from nuclear power stations, have caused most concern. Nuclear blasts have contaminated the test sites with long-lived radioactive isotopes, and the nuclear 'fallout' has been deposited as fine dust across the ecosystems of the world, where the isotopes have been incorporated in food webs. Despite the possibilities that radioactive strontium from the fallout might have contributed to bone cancer in American children, and radioactive iodine to thyroid cancer in eskimos (iodine→lichens→caribou→eskimos), convincing statistical evidence for these effects is lacking. The disposal of high level nuclear waste from power stations, however, is a sensitive political issue. The waste has to be transported without risk, and stored underground in geological strata which are unlikely to be disturbed.

Central to the debate about nuclear power stations is the possible effect of small doses of radiation in promoting cancer. Whilst elevated levels of leukaemia have been found in children near the nuclear reprocessing plants at Dounreay in northern Scotland and Sellafield in Cumbria, it is possible that viruses, rather than radiation, may be involved.

The explosion of a nuclear reactor at Chernobyl, in the Ukraine, in April 1986 showed that even a small accidental release of radioactivity can have lasting ecological effects at some distance. The dust cloud passed over Scandinavia and many of the isotopes were deposited by rainfall over Cumbria and North Wales. Radioactive caesium was picked up by plant roots and transferred to sheep and cows grazing the grass. As a result the areas suffered an embargo on the sale of mutton and milk. The caesium seems to be recycled tenaciously by the grasses and so some of the bans on mutton sales are still (1989) enforced.

CHLOROFLUOROCARBONS

These gases are manufactured as propellants in aerosol cans, for the heat exchange units of refrigerators, and for blowing polystyrene foam in the manufacture of packaging. Apart from their contribution to the 'greenhouse effect' (p. 112), they appear capable of reducing the thickness of the ozone layer in the stratosphere. The recent (1985) discovery of temporary holes in the ozone layer over both poles has hastened moves to reduce the production of CFCs by the developed countries. For the ozone layer traps much of the 'ultraviolet B' radiation from the sun. A thinner ozone layer would allow more ultraviolet light to pass through it, increasing the likelihood of mutations, skin cancers and cataracts.

The output of CFCs is being reduced by international agreement, but it may be too late. It will take some time for alternative compounds to be manufactured on a large scale, and the problem has arisen at a time when refrigerators are being adopted by the less developed countries. The media campaigns against CFCs have highlighted the fragility of the planet and the need for greater international co-operation and altruism in dealing with pollutants.

In many industrialised countries, the decline of wildlife in the face of pollution and other threats has been rather dramatic. The recent science of 'Nature Conservation' has developed, devoted to the preservation of semi-natural ecosystems and individual species.

Nature conservation

Pollution is not the only threat to semi-natural ecosystems and wild species of animals and plants. Humans, through deforestation and cultivation, have reduced semi-natural ecosystems to fragments of their former areas. Many species are teetering on the brink of extinction. In the face of these pressures, societies and governments throughout the world have 'conserved' pieces of landscape in their 'natural state' as nature reserves and national parks (figure 92).

The word 'conservation' refers to the wise use of natural resources for the benefit of mankind. Wild species of animals and plants constitute an important natural resource. For example, conservationists would argue that instead of overfishing to extinction wild populations of whales, anchovies and herrings, the catches should be reduced. Only the 'interest' should be taken but the capital should be left intact (see p 75). In that way the food resource might be continuously available for thousands of years.

Conservation, as distinct from preservation, is an active not a passive process. One cannot merely put a fence around a nature reserve and expect the vegetation to remain the same and the rare species to persist for ever. If it would benefit the nature reserve in the long run, the warden might allow public access or even promote it, set fire to the vegetation, fell some trees or shoot some of the animals. If one of the rarer plants or animals on a reserve shows signs of dying out, its numbers may be maintained by creating new habitats for it, cross pollinating it, or rearing it in captivity and then releasing it back into the wild.

In Britain many organisations have sprouted up to promote the conservation of wildlife. The Nature Conservancy Council, (NCC), funded by the government, has over 200 National Nature Reserves. Its main role is to conserve a representative sample of the main types of ecosystem in Britain. Yet it stimulates and finances surveys and research on the whole of British wildlife. The Forestry Commission owns over 400 000 ha, much of it planted with alien conifers. Despite this it has recently paid considerable attention to the amenity planting of deciduous trees and to conserving the variety of

FIGURE 92
Yosemite National Park, California, one of the first areas in the world to be set aside for public use and recreation, in 1864. (U.S. National Parks Service.)

wildlife in its forests. The National Trust also has 200 000 ha of wild open space. The Royal Society for the Protection of Birds, (RSPB), controls over sixty reserves, which are havens for many other organisms besides birds. The Country Naturalists Trusts have each at least thirty reserves. Apart from these, many Universities and Field Study Centres administer their own teaching and research sites.

Is all this effort worthwhile? What justification is there for conserving areas of semi-natural wilderness? Should we not merely allow the destruction of nature to take its course?

THE CASE FOR NATURE CONSERVATION

Some of the arguments in favour of nature conservation are rather esoteric and nebulous and might carry little weight in a dispute between conservationists on the one hand and industrial giants on the other. Nevertheless professional ecologists, amateur naturalists, country-lovers, anglers and holidaymakers in Britain form an articulate and emotionally involved lobby which amounts to a large pressure group in favour of the conservation of what remains of our semi-natural ecosystems.

An example of the sorts of conflict which arise is the Upper Teesdale controversy. In 1965, ICI planned to build a new fertiliser plant at Billingham-on-Tees, with the creation of 4000 new jobs. The factory required a water reservoir on the Tees, and the site chosen was Upper Teesdale, with its pocket of rare arctic-alpine plant species at the southern ends of their geographical ranges. The Botanical Society of the British Isles opposed the plan at the public enquiry, calling the site the 'Westminster Abbey of British Botany'. When the inquiry came down against them, the botanists took their case to the House of Lords, but lost it on a free vote despite intensive lobbying. ICI built the reservoir (figure 93) but donated £100 000 for research into the plants on the site. Some plants in the community were flooded when the reservoir was created, and the remainder were expected to be wiped out by the presence nearby of a large sheet of water. Yet all the important arctic-alpine species still survive there.

Some of the arguments which can be advanced for the preservation of semi-natural ecosystems and individual species are as follows.

(a) Moral or ethical. Wild species have a right to exist, but no power over their own survival. Humans, as the dominant organisms on the planet, are morally responsible for the survival of the species under their care.

(b) Aesthetic. Animals and plants look attractive, and observing them gives many of us considerable pleasure. They are the basis of much of our culture, our art and music. They are worth preserving because of our emotional response to them.

FIGURE 93

Upper Teesdale. The Cow Green dam with its partially filled reservoir. The river Tees is in the foreground. The dam was completed in 1970 and the lake now covers 312 ha. (ICI).

(c) Psychological. In intensively cultivated, industrialised and urbanised countries, the presence of open green spaces where humans can relax in a peaceful and secluded atmosphere is essential to their psychological well being and hence the sanity of society. The countryside is the habitat to which our ancestors have been adapted by natural selection. Most humans have only abandoned the countryside during the last few generations. Although it is impossible to prove, the need of many individuals to grow pot plants in their houses and to cultivate their own gardens may be an expression of the psychological need to feel closer to nature in an urban society.

(d) Outdoor laboratories. Firstly, ecological investigation has a value in its own right as a field of intellectual enquiry and areas of semi-natural ecosystems ought to be set aside for its study. Secondly, pure research may have unexpected value to man. For instance, if a wild species becomes a pest of agricultural crops, and its natural history is already well known, a biological control agent should be easy to find and should still exist. Thirdly, the measurements and descriptions made by scientists on semi-natural ecosystems form the basis of our understanding of nature. In the long run this understanding will allow the repercussions of human activity on the landscape to be predicted with much greater certainty.

(e) Pollution monitors. The reactions of populations of wild species to environmental stresses may provide us with an early warning of pollutants which may be having sub-lethal effects on human populations. For example, the decline of grebes, peregrine falcons and other species at the tops of food chains first drew our attention to the toxicity of DDT. Wild species should therefore be preserved as potential pollution monitors, working in much the same way as the canaries which signalled to miners the presence of toxic gases.

(f) Balance of nature. According to this argument, the destruction of areas of wilderness or indeed, any interference with nature is unwarranted because it might alter the 'balance of nature'. This balance can be upset in two main ways. Firstly, the disturbance might trigger off a chain reaction and cause ecological damage far from the site which was disturbed. Secondly, the removal of a particular species might alter the balance between the remaining species in the ecoystem. In either case the repercussions on the human population may be undesirable.

The creation of large deserts has sometimes begun with small scale disturbances. In Africa and India, overgrazing of small areas by sheep, cattle and goats has been followed by a reduction in soil organic matter, the death of the vegetation through drought, and the burial of healthy vegetation nearby under wind-blown sand. Once this process has started the desert margin may spread at a rate of several kilometres a year.

Even species which are apparently insignificant may be vital to the proper functioning of an ecosystem. In a salt marsh in the south eastern United States there are only about eight ribbed mussels (*Modiolus* spp.) m^{-2}, but they are very important to the ecosystem. When they are covered with water, at each high tide, they filter out organisms, and particles which are rich in phosphorus. They feed on them, and deposit much of the phosphorus in excreta on the surface of the mud. In this way they trap the important nutrient element phosphorus, and make it available to the other plants and animals in the ecosystem, instead of allowing it to disappear when the tide retreats. If the mussels were eliminated, the primary production of the marsh would declne and its vegetation might change dramatically.

The disappearance of predators may allow their prey to flourish and reach pest proportions. One reason for the increase in the wild rabbit population in Britain in the middle of the eighteenth century was that gamekeepers, trying to increase their stocks of pheasants and partridge, shot and eliminated many birds of prey. These birds also fed on rabbits, which freed from some of their predators, underwent a population explosion.

The 'balance of nature' argument for nature conservation is rather a neurotic one. We should not disturb ecosystems much, and we should not allow any species to die out. Otherwise there might be unexpected and deleterious effects on those ecosystems which support man.

(g) Productive species. Many wild organisms may be particularly efficient energy converters or productive sources of food. We ought to preserve them, and the ecosystems which support them, in case we need them as food sources in future. Fish stocks in the sea are capable of supplying a great deal of the world's protein for a long time if they are harvested with care. Many wild herbivores on land can use poor pastures more efficiently than introduced cattle. Red deer, for instance, are well suited to the rugged environment of the Scottish highlands and will probably be farmed there for meat on an increasing scale.

(h) Plants which contain useful compounds. Many flowering plants may contain compounds useful to man, but only 5000 species out of 250 000 have been screened intensively from this point of view. Many potent compounds come from wild and cultivated plants. Digitoxin, used in medicine to speed up the heart, is extracted from foxglove (*Digitalis purpurea*) leaves. Morphine, used as a pain killer, comes from poppy fruits. Atropine, which dilates the pupil of the eye, is present in all parts of the deadly nightshade plant. Hormones like oestrogens, progesterone and cortisone can be made from compounds extracted from soyabeans (*Glycine soja*) or the rhizomes of yams (*Dioscorea* spp.). Foodstuffs, oils, alcoholic drinks, rubber, resins, fibres and dyes can all be found in wild plants. If individual species are allowed to die out, we may lose their potential uses for ever.

(i) Genetic conservation. If a species becomes extinct, its genes, which may be unique, may be lost for all time and cannot be used to breed better races of crop plant or animal. This probably applies more to plants than animals.

Cultivated strains of wheats, for instance, have benefited from time to time by interbreeding with wild species. The wild relatives may possess characteristics like hardiness, resistance to lodging (falling over when the wind blows), and fungal resistance, which the cultivated strains lack or have lost. If the wild wheats die out in their intensively cultivated habitats in the Middle East, this sort of improvement will not be possible. At the wheat seed banks which have been set up all over Europe, e.g. at Bari in Italy, thousands of wild wheats and cultivated strains are grown, and their seeds stored under optimum conditions, in case they are needed in future.

(j) Economic. There are sound financial reasons for conserving wild areas and particular species in many areas of the world. Kenya, for example, gains a third of its foreign exchange currency from tourism, which is based on the attraction of its National Parks, full of big game animals. Grouse moors, pheasant shooting, salmon fishing and trout streams in Britain are all lucrative enterprises which depend on the conservation of wild ecosystems. Many of the wilder regions of Britain, Europe and the United States are economically depressed. Yet tourists seeking peace, solitude and wildlife contribute a great deal to the economy of the areas and create many new jobs. This, because it is expressed in hard financial terms, is possibly the argument for nature conservation which carries most weight in politics.

Conservation policies
Government policies towards nature conservation vary widely. Costa Rica, for example, is making strenuous efforts to conserve what is left of its forests, whereas Brazil at present (1989) acquiesces to large-scale forest destruction. There are many independent organisations which promote nature conservation at an international or national level (figure 94). They range from the IUCN and the World Wildlife Fund for Nature, staffed by diplomats and with international backing and credibility, to schools and local Naturalists' trusts which conserve small areas of landscape. Some of this work is devoted to the conservation of large areas of threatened habitats, and some to preventing the extinctions of rare species.

The prime example of a threatened habitat is the tropical rainforest. The destruction of this fragile biome, with its astonishing diversity of plant and animal species, is proceeding at a dramatic pace. Forty-four per cent of its original area has disappeared, and further destruction takes place at the rate of 10–20 million hectares a year. Biologists have not had time to find out how tropical rainforest ecosystems work, or to identify all the species.

The protection of significantly large areas of tropical forest seems the only way forward. In small reserves the population sizes of species are so small, especially in the tropics (p. 58), that extinction is quite likely. The results of the 'Minimum Critical Size Project', in the Amazon basin near Manaus, Brazil, suggest that it might be desirable to preserve whole areas of at least 70 000 ha. This huge experiment allows biologists to monitor the process of extinction and species change in forest reserves ranging from 1–10 000 ha in patches of habitat which have been isolated from once-continuous forest.

The whole question of the best sizes and shapes for nature reserves has come under critical scrutiny. When a reserve is part of a continuous stretch of landscape, the rate at which it gains new species from nearby should roughly equal the rate at which its existing species becomes extinct. Isolate the reserve, however, and the immigration rate should fall, whilst the extinction rate should remain at least the same. It follows that the rarer species in a reserve may need some protection, management, or habitat

FIGURE 94
Some important organisations involved in nature conservation.

Organisation	Functions
International Union for the Conservation of Nature and Natural Resources (IUCN)	World-wide co-ordination of action by governments on conservation issues.
World Wildlife Fund for Nature (WWF)	International fund-raising for conservation projects.
Nature Conservancy Council (NCC)	U.K. government's conservation agency. Manages about 200 National Nature Reserves (150 000 ha). Designates Sites of Special Interest (S.S.S.I.s) —4000 areas (1.3 million ha) on which development may be limited under the Wildlife and Countryside Act (1981).
Forestry Commission	Government agency. Owns and manages 1.3 million ha of British forests.
Countryside Commission	Government agency. Policy on National Parks, Areas of Outstanding Natural Beauty, public access to countryside.
Royal Society for the Protection of Birds (RSPB)	Voluntary. Eighty reserves (40 000 ha) and active in protecting rare species.
National Trust	Voluntary. Owns 1% of land in England and Wales. Manages many important sites.
Royal Society for Nature Conservation (RSNC)	Umbrella body for 42 County Naturalists' Trusts, which manage about 1500 local reserves (170 000 ha).

creation if they are to survive, especially in small reserves which may only support low population sizes.

All individual species are worth conserving, unless they are pests. This is especially so when they have some symbolic importance which attracts media interest in conservation, such as the bald eagle in the United States or the giant panda in China. Threatened species are listed in the *Red Data Books* of the World Wildlife Fund for Nature and many individual nations. There have been numerous successes. For example, only fifty Hawaiian geese (Ne-nes) existed in 1950, but over 1600 have since been bred in captivity at the Wildfowl Trust, Slimbridge, Gloucestershire, and released onto Hawaii. Pere David's deer were extinct in the wild in their native China in 1865, but eighteen were imported to Woburn Abbey, Bedfordshire in 1900.

From this only surviving stock enough animals were bred to re-introduce the deer to its native China in 1985.

The large blue butterfly (*Lycaena machaaon*) seems to have become extinct in Britain in 1979 for a variety of reasons, including habitat destruction, over-collecting, freak droughts and the elimination of the ants on which its life-cycle depends. It only occurs in short grassland in the presence of the plant wild thyme (*Thymus pulegioides*) and the ant *Myrmica sabuletii*. Its caterpillars eat thyme and are adopted by the ant colony, in which they pass the winter. When the virus disease myxomatosis largely eliminated rabbits in 1953–5, tall grasses grew up in many grasslands, eliminating the ants. The butterfly was successfully re-introduced into Britain from Sweden in 1987 into properly-managed habitats. Was the money spent on this project worthwhile? It may have been justified from the point of view of publicity, but some would argue that the cash might have been better spent in conserving whole habitats.

In the 1960s ecologists were suggesting that the more species an area supports, the more 'stable' the community might be, in the sense that the population fluctuations of the constituent species should be reduced. In fact, rather than 'diversity begets stability', it seems more likely than 'stability begets diversity', in the sense that those habitats with the smallest environmental fluctuations throughout the year support most species. Nevertheless, computer models of food webs suggest that population fluctuations decrease as the number of trophic levels increases, and in the presence of omnivores. This implies that some species might be almost indispensable cogs in the machinery of nature. With thousands of species becoming extinct each year, however, we have little time to find out!

10 | *Practical section*

You must perform an ecological project or some class experiments. These will improve your understanding and complement the theory teaching. This section provides some advice about projects, and describes some investigations suitable either for class experiments or for project work. A list of suitable practical texts and identification books is provided in the 'REFERENCES PRACTICAL' section on pp. 175–177. When these texts are mentioned in this section, they are given author and date, for example, Bailey (1981). Otherwise the references to individual articles are given in full.

Projects

This section provides firstly, general advice on how to select and carry out your project, and secondly, a list of ideas for ecological projects with some references. For more detailed advice, I would recommend M. Wedgwood (1987) *Tackling Biology*

Projects (Macmillan/Institute of Biology), and for practical techniques, G. M. Williams (1987) *Techniques and Fieldwork in Ecology* (Bell and Hyman). Slingsby and Cook (1986) discuss the planning of field investigations and Dowdeswell (1984) provides an introduction to fieldwork in specific habitats.

ADVICE ON PROJECTS

The secret of a successful ecological project is to select a topic which is both suitable and interesting. An infinite number of different projects is possible. Most of them are not worth attempting. Be particularly wary of choosing a project which (**a**) merely proves the obvious, (**b**) involves much identification or (**c**) requires the collection of many environmental measurements. In particular, avoid investigations of whole animal or plant communities. Identification of animals and plants down to species level can be difficult and time-consuming. Instead, concentrate on the distributions of five or six easily identified species. Remember that the best way to unravel cause and effect in ecology is to carry out *experiments* in field or laboratory.

HOW TO PLAN AND EXECUTE YOUR PROJECT

Search for a topic which really interests you.

(a) Find a phenomenon, e.g. slugs congregate beneath a fallen tree trunk during the day.

(b) Examine the site carefully. The more time that you spend on the initial survey, the more you will learn and the more ideas will dawn on you.

(c) Decide on a particular *hypothesis* to be tested. An hypothesis is a tentative explanation of the phenomenon, which can be tested by experiment or observation, for example, slugs congregate beneath the tree trunk because they move towards areas with the highest relative humidity.

(d) Decide how your data are to be collected. What method is to be used to sample the organisms? What measurements and counts are to be made? Do not be too grandiose in your schemes. Beginners greatly overestimate what is possible in the time available. The analysis of the data and the production of the project report will take at least as long as the practical work.

(e) At this stage you should make a list of all the apparatus which you require and check that it is available.

(f) Carry out a pilot experiment or make a trial set of observations. This will indicate the snags in the experimental technique, suggest extra measurements which might have to be made, and enable you to predict how long the practical work will take. Above all, it will indicate if the project is viable or not.

(g) Consider how many samples should be collected, or how many experimental treatments and replicates should be set up. In this respect one major consideration will be time. Other limitations may be bench space and the availability of materials and equipment. Bear in mind that some replicates may fail and you should increase the number to allow for natural wastage.

(h) Even before you start the project, think about how you will present the data in your report. Will negative results be valuable? For example, if the slugs showed no preference, in experiments, for a particular relative humidity, could you still write up the project in an interesting manner?

(i) Before you start an experiment consider whether the data can be analysed statistically, for instance by performing a significance test. You may have to consult a teacher and a statistics text to determine which test is most appropriate.

The sort of test which is to be applied affects the planning of an experiment. To compare two sets of data, say the growth of duckweed under two different nutrient regimes, each treatment needs seven or eight beakers if a Mann-Whitney U test is to be usefully employed. Each treatment would need a minimum of three beakers for a conventional *t*-test. As the sample size increases, so does the precision and value of the test.

(j) When you are satisfied on all these points, you can begin to collect the data!

IDEAS FOR PROJECTS

Most projects are investigations of phenomena which a student notices near home, at school, or on a field course. The projects listed here are those which can be performed almost anywhere. I have largely excluded projects which can only be performed in one sort of habitat. The list emphasises experimental rather than descriptive projects because I firmly believe that they are more valuable.

1. Factors affecting distribution patterns of different animal species, e.g. planarians, shrimps, barnacles, mussels.
2. Microhabitats of different urban fern species.
3. Comparison of microhabitats and behaviour of two species of woodlice in field and laboratory. (e.g. J. M. Dangerfield *et al.*, *Journal of Biological Education* 21 (4), 251–258, 1987).
4. Ants; various descriptive projects. (G. J. Skinner 'The use of ants in field work'

Journal of Biological Education 22 (2), 99–106, 1988).
5. Effects of trampling and burial on seed germination and burial in the greater plantain (*Plantago major*).
6. Effect on moss growth of varying nutrient concentrations in the substrate.
7. Comparison of soil organisms in two or more habitats (see p. 153).
8. Mosses or protoctists on north/south sides of walls or different levels up a salt marsh or rocky shore. Distribution pattern in relation to desiccation tolerance (p. 161).
9. Distribution patterns of nettles in relation to soil phosphate (e.g. Monger 1986).
10. The effects of different water regimes on plant growth (e.g. Monger 1986).
11. Soil pH and nutrient concentration around the bases of coniferous trees; effect on earthworm population (see p. 158).
12. Pattern of earthworm abundance and soil humus levels with distance from a tree trunk.
13. Location of nitrate reductase and nitrite reductase enzymes in plants of different species (P. W. Freeland, *Problems in Practical Advanced Level Biology*. Hodder and Stoughton 1985).
14. Light compensation points of sun and shade plants.

15. Light intensity and the distribution pattern of woodland herbs (p. 157).
16. Rate of fall of, and distance travelled by, winged seeds in relation to their structure (Roberts and King 1987, *Investigation 24.11*).
17. Inter-specific interaction between clover and grass; addition of fertiliser (Monger 1986).
18. Inter-specific competition amongst two duckweed species (e.g. Monger 1986 and p. 158).
19. Competitive relationships between plants, assessed in the field (see W. J. Sutherland and A. R. Watkinson, 'Ecological studies of plants'. *Journal of Biological Education* 22 (1), 31–36, 1988).
20. Pot experiments on competition between a crop plant and a weed.
21. Competition between different genotypes of fruit flies (*Drosophila*) (e.g. Monger 1986).
22. Life histories of insects isolated from galls.
23. Comparison of diets of different species of snails and slugs by faecal analysis (p. 159).
24. Pollination in Lords and Ladies (*Arum maculatum*) (D. H. T. Jones, *Journal of Biological Education* 11, 253–260, 1977).
25. Pollination; flower constancy of pollinating insects; pollen identification from insects; flower preferences in relation to tongue length; role of flies in pollen-eating and pollination. (see p. 160); also T. J. Aston, 'Plant–pollinator interactions; a rich area for study' *Journal of Biological Education* 21 (4), 267–274, 1987.
26. Populations of domestic pest insects in households (see B. Turner, *Journal of Biological Education* 22 (3), 183–188, 1988).
27. Growth curve in a population of bacterium or protoctist assessed with a colorimeter or haemocytometer (p. 163).
28. Population dynamics of plants in permanent quadrats; death rates of seedlings and mature plants (p. 165).
29. Population regulation in the water flea (*Daphnia*) (p. 163).
30. Birth and death rates of different leaves on the same plant or leaves on plants from contrasting habitats.
31. The effects of ladybirds, ladybird larvae or hoverfly larvae on greenfly populations (p. 163).
32. Predation rates of damselfly or mayfly nymph on water fleas (p. 164).
33. Life tables for the holly leaf miner (*Phytomyza ilicis*) and its parasitic insects (Lewis and Taylor 1967, Monger 1986).
34. Interaction between spear thistles and gall flies (M. Redfern, *Journal of Biological*

Education 22 (2), 88–90, 1988).
35. Food web for a pond by examining gut contents and putting organisms together in pairs (Monger 1986).
36. Successional sequences of lichens and mosses on gravestones.
37. Succession of animals in dung pats.
38. Succession in laboratory ecosystems or in washing-up bowls of different sizes.
39. Decomposition rates of leaves in mesh bags of different mesh sizes, under trees of the same or different species (p. 157).
40. Energy budgets for stick insects of various ages (see p. 166).
41. Distribution patterns of lichens at various distances from a potential pollution source, e.g. a power station (p. 170).
42. Effects of simulated acid rain on seedlings (e.g. G. Fleet *et al.* 'Acid rain in the classroom: a student research project'. *Journal of Biological Education* 21 (3), 156–158, 1987).
43. Testing for salt tolerance or lead tolerance in races of grasses collected from roadsides; comparison with normal races (p. 168).
44. Growth of duckweed (*Lemna* spp.) at different levels of added nitrate and/or phosphate to simulate eutrophication (p. 171).
45. Growth of duckweed (*Lemna*

spp.) at different levels of added detergent to simulate pollution.
46. The effects of various treatments on the re-vegetation of trampled paths.
47. Biochemical oxygen demand (BOD) and fauna in rivers at different distances from a pollution source (p. 169).

Investigations

In the next few pages are listed several investigations which are suitable for classes or individual project work. Naturally, most ecological investigations are determined by local conditions. In selecting these ideas from the large number available I have applied four main criteria:

(*i*) Investigations should illustrate ecological principles.

(*ii*) They should, as far as possible, be experimental rather than descriptive.

(*iii*) The organisms which are employed should be widely available in schools and colleges.

(*iv*) The investigations should, as far as possible, fit into practical lessons and be carried out in the laboratory or in school grounds.

INVESTIGATION I

Extraction and identification of organisms in leaf litter

Requirements
Soil or leaf litter, collected from

(a) different depths of the same soil,

  (b) beneath different tree species, or

  (c) from completely different habitats.

Tullgren funnel (filter funnel, gauze, lamp, beaker)

Baermann funnel (filter funnel, muslin, lamp, beaker)

30% methanol

● Set up the funnels (figure 95).
● Identify the extracted organisms as far as possible, using the keys in Leadley-Brown (1978) Lewis and Taylor (1967) or Paviour-Smith and Whittaker (1971).
● Tabulate the results.

  What major differences exist between the faunas in the two samples? Might these

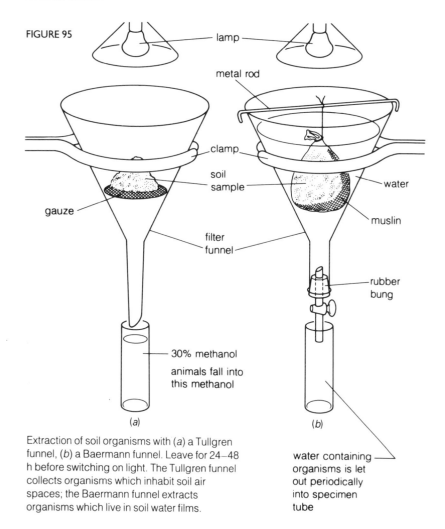

FIGURE 95

Extraction of soil organisms with (a) a Tullgren funnel, (b) a Baermann funnel. Leave for 24–48 h before switching on light. The Tullgren funnel collects organisms which inhabit soil air spaces; the Baermann funnel extracts organisms which live in soil water films.

water containing organisms is let out periodically into specimen tube differences be attributable to the quality or quantity of food available? How are these organisms adapted to live in the micro-environment in which they are found?

INVESTIGATION 2

The roles of different organisms in breaking down leaf litter

Requirements
Leaves of one or more species of deciduous tree in autumn
Nylon mesh bags

- Enclose leaf discs of known surface area or mass in nylon mesh bags.
- Apply various treatments, ensuring that four repeat samples (replicates) of each treatment are set up. Treatments might include (a) various mesh sizes, (b) different depths or (c) burial beneath canopies of different tree species.
- Remove them at intervals of a month, reweigh or remeasure them, and replace them. A mesh size of several mm allows all soil micro-organisms and invertebrates to enter. Sizes such as 1 mm exclude earthworms but not mites, springtails or micro-organisms. Very fine mesh keeps out all invertebrates but allows fungi and bacteria to grow.
- Relate the experimental results to the natural rate of decomposition of the leaves of the same species. Do the leaves

of different species break down at different rates?

What is the relative importance of (a) earthworms, (b) other detritivores, and (c) fungi and bacteria, in the breakdown of leaf litter? What influence does the depth of burial have on the decomposition of leaves?

INVESTIGATION 3

Light intensity and the distribution of woodland herbs

Requirements
Deciduous woodland with several herb species
Photocell or photographic light meter
Bamboo canes
Pack of playing cards

- Perform this exercise on a uniformly sunny or overcast day.
- Select a wood with some gaps in the tree canopy.
- Pace out two coordinates at right angles. Work out the x and y coordinates of random sample points by using the pack of cards. Cut the pack four times to establish each sample point. The first two figures (counting a 10 as zero, an ace as one, and ignoring the royal cards) represent the x coordinate on a 100 m axis, and the last two figures represent the y coordinate.
- Pace out each point and mark it with a bamboo cane. Record which of the five or six easily

identified herb species included in the study are present within 1 m of the sample point. Then find the next point and repeat the procedure.

- Measure the light intensity 1 m from the ground at each cane in rapid succession. Use the same light meter for all the readings. Repeat this round of light measurements once or twice.
- Calculate the average light intensity at each point.
- Draw histograms of (a) frequency of sites with light intensity and, (b) percentage of sites with a certain species at each light intensity. Test if the presence of a species is independent of light intensity with a 2×2 contingency table (Bailey 1981, Campbell 1974).

  Do species prefer different light intensities? What do you think prevents them from growing at (a) higher light intensities and (b) lower light intensities? Is light intensity the only factor involved? How would you assess the importance of other factors?

INVESTIGATION 4

Intra and interspecific competition in duckweeds (*Lemna* spp.)

Requirements
Duckweed species
Beakers or plastic cartons in greenhouse

- Grow the same species at four or five densities (e.g. 4, 8, 16, 32, 64) in beakers or plastic cartons of pond water.
- Grow two different species at the same total density but in different relative proportions.
- Replicate each treatment four or five times. Randomise the beakers on the greenhouse bench.
- Count the living and dead leaves in each beaker each week.
- Alternatively, if enough materials are available, set up this whole experiment four times over. At 10-day intervals harvest the species from a single group of beakers and determine their dry masses.
- Graph your data.
  What can you deduce about (a) competition between individuals of the same species and (b) competition between different species? (See also Monger 1986).

INVESTIGATION 5

The distribution of earthworms in relation to soil conditions

Requirements
Watering can or spade
1% formalin, or 2% potassium permanganate, or soapy water
Quadrat
Electric pH meter
Balance
Bunsen
Tripod and gauze

- Compare earthworm populations beneath different tree species or in different habitats.
- Extract earthworms by sprinkling a known area with one of the solutions listed above and collecting the worms which emerge over 20 minutes, or by digging up a certain volume of the soil and sorting it by hand. The former method is more suitable for species which live at depth, such as *Lumbricus terrestris*. The latter method is most efficient for the smaller surface dwelling species.
- Consult Chinery (1977) for identification.
- Count the worms, weigh them individually. Measure the resting lengths of the individuals. Draw frequency distributions of length and fresh mass.
- Suggest reasons for the differences between habitats in the masses or length distributions of the worms.
- Compare the pH and humus content of the soils from which worms were collected. Humus content is estimated by the loss in mass of dry soil heated strongly. pH is determined by using the electric pH meter.

  Can you relate the abundance or species distribution of earthworms to pH or food supply?

INVESTIGATION 6

Comparison of the diets of slug or snail species

Requirements

Slugs and snails collected from the same habitat on a moist day
Plastic containers
Cavity slides
Microscope
Coverslips

- Identify the species (Beedham 1973, Chinery 1977, Kerney and Cameron 1979).
- Place each individual in a separate plastic container for 24 h.
- Sample at least four individuals from each species.
- Collect and store the faecal pellets from each slug in a labelled tube of acetic alcohol. Examine droppings soon after collection. At this time green chlorophyllous tissue is still distinguishable from brown dead tissue. Tease out a pellet in water on a cavity slide. Examine it under a monocular microscope.
- Estimate by eye the proportion of the solid matter in the field of view which consists of:
  (*a*) living (green) flowering plant tissue
  (*b*) dead (brown) flowering plant tissue
  (*c*) living moss
  (*d*) dead moss
  (*e*) fungus
  (*f*) pollen grains
  (*g*) arthropods
  (*h*) the rest.

- Incidentally, the leaves of many herbs are digested and are absent from the pellets. Epidermal strips of grasses may pass through the gut unchanged.
- Tabulate the results. Calculate the means and standard errors (Bailey 1959, Campbell 1974) for each constituent for each species.

  Do the diets of species differ? Do the species compete with one another for food? Or is their food superabundant?

INVESTIGATION 7

Flower preferences and foraging strategy of honey bees and bumble bees

Requirements
Garden or semi-natural habitat with many flowers in bloom
Warm summer day
Conical flask and bung
Cotton wool and ether
Compound microscope, cavity slide, coverslip

For identification of bee species see Chinery (1977), Lewis and Taylor (1967) or Proctor and Yeo (1964).

Experiment One
- Follow individual bees and note (a) their methods of obtaining pollen and nectar, (b) the successive flowers visited, (c) the direction in which the bee moves between the flowers in the inflorescence and (d) the direction of the wind.

  Is there much variation in flower preference and foraging strategy within a species? Compare the foraging methods of different bee species. Which is most efficient at obtaining food? Are some flowers not being visited by bees? Why not?

Experiment Two
- Check the range of plant species visited by each bee species (experiment one). Different plants produce pollen grains which are distinctive in size, shape and sculpturing (see Hodges 1964).
- Catch a bee by placing over it the opening of a small conical flask. Close the flask with a bung with an ether-filled pad of cotton wool beneath.
- Remove the etherised bee and extract the contents of a pollen basket with forceps (care!).
- Place the pollen in water on a cavity slide. Tease it out with mounted needles. Add a cover slip and examine the pollen under the high power of a microscope. Record the number of pollen grains of each type within the field of view. Examine pollen from flowers in the foraging area. Compare your results with those from experiment one.

  Which method is more reliable? How does flower constancy differ amongst bees of (a) the same and (b) different species?

Experiment Three

● Relate the average tongue length of each species to the average corolla tube length in the flowers that they visit (Lewis and Taylor 1967). The tongue lengths must be determined on a *small* sample of killed workers. The length of the corolla tubes of ten flowers of each species visited must be measured and averaged.

How much influence does corolla tube length have on the range of flowers favoured by each bee species?

INVESTIGATION 8

The distribution patterns of seaweeds or mosses in relation to relative humidity

Requirements

Adjacent habitats which support different seaweeds or moss species, e.g. rocky shore, or different aspects of a wall, or the top of a wall and long grass
Humistors, cobalt chloride paper or Piche evaporimeters
Spring balance
Containers with anyhydrous calcium chloride or silica gel

● Map the distribution of each species.

Experiment One

● Relative humidities can be compared with *(a)* humistors, small sensitive devices which work by electrical resistance and provide a direct read-out of relative humidity on a dial, or *(b)* the time taken for anyhydrous cobalt chloride paper to go pink.

Experiment Two

● Compare the evaporating power of different micro-environments

FIGURE 96

Piche-type evaporimeter which can be used to compare the rates of evaporation of water in different environments. The rate at which the meniscus travels along the capillary tube is recorded.

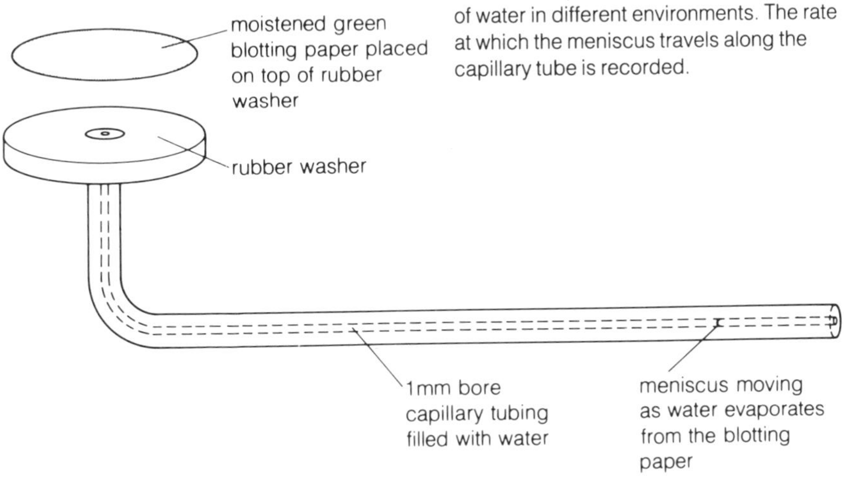

moistened green blotting paper placed on top of rubber washer

rubber washer

1mm bore capillary tubing filled with water

meniscus moving as water evaporates from the blotting paper

with Piche evaporimeters (figure 96).

- Calculate for each site the volume of water which evaporates per unit time.

Experiment Three

- Seaweeds, mosses and liverworts can be placed in atmospheres of various relative humidities. Measure their masses at intervals with a spring or electrical balance. Different species lose water at different rates (Evans and Hardy 1970).

Experiment Four

- Measure the resistance of mosses or liverworts to desiccation by exposing them to anhydrous calcium chloride or silica gel in closed containers. Examine a leaf sample at intervals under high power. Dead cells do not swell when the leaves are soaked in water for a few minutes. Plot the percentage of dead cells against time for different species.
- Relate the distributions of the species in nature to (*a*) the relative humidity or evaporation rate of the environment and (*b*) the rate of water loss and desiccation resistance of the species.

INVESTIGATION 9

Mark–release–recapture estimates of population size

Requirements

Large populations of a small mobile animal, without strong flight,

e.g. froghoppers, grasshoppers, pond skaters, centipedes, woodlice, snails, or even voles and shrews caught in Longworth traps
Paint brushes
Oil or cellulose based paint

- Compare the population density of a species in different microhabitats or at different times of year. Alternatively, compare the numbers of two closely related species in the same habitat. Proceed as follows:
- Mark out a well defined area.
- Capture as many individuals as possible and mark them with paint in a standard manner. For insects a spot on the thorax is usually sufficient.
- Discard damaged individuals.
- Release the marked animals throughout the area. Allow enough time for these individuals to mix thoroughly with the unmarked individuals of the same species.
- Recapture as many individuals as possible. Count the total number and the number of these which have already been marked. The population size of the species is MC/R, where M is the number of marked animals released, C is the total number of animals recaptured and R is the number of recaptured individuals which had already been marked.
- Make a list of the assumptions on which this method is based.

Are you satisfied that your estimates of population size are reliable? Suggest explanations for the differences in population density you find. How might you test your explanations experimentally?

This investigation can also be performed on populations of known size kept in trays in the laboratory (e.g. the flour beetle *Tribolium*, Wratten and Fry, 1980; Roberts and King, 1987), or represented by poppet beads (e.g. Monger, 1986). The principle of the method is most easily demonstrated by sampling a certain number of individuals from a population of fruit flies (*Drosophila*) which contains a known number of an easily recognised mutant, representing individuals which have been 'marked' (E. Coulton, pers. comm.).

INVESTIGATION 10

Population growth; multiplication of yeast cells (see also Roberts and King (1987) *Investigation 32.5*)

Requirements
Microscope
Haemocytometer
Graduated cylinder (50 cm³)
Conical flask (250 cm³)
Cotton wool
Pipette
Graph paper
Nutrient medium for yeast (e.g. 50 cm³ 2% sucrose)
Suspension of brewer's yeast.

● Pour the nutrient solution into the conical flask and add a drop of yeast suspension. Plug the flask with cotton wool and leave it in a warm cupboard, oven or water bath.

● Using a haemocytometer, take cell counts twice daily over at least three days. On each occasion take two counts and average them.

● Plot population size and the logarithm of population size against time. Determine the doubling time for the yeast population from the semi-logarithmic plot.

A similar experiment on the protoctist *Chlorella* is described in Monger (1986), Investigation 28C. A more sophisticated experiment on population regulation in the water flea, *Daphnia*, examines the effects of density-independent and density-dependent mortality (Monger 1986, Investigation 28D). There are three treatments. In one, the population is allowed to grow unchecked, in a second, a random number of *Daphnia* is removed each week, and in the third, more *Daphnia* are removed with increasing population size.

INVESTIGATION 11

The effect of predators on aphid populations

Requirements
Broad bean plants in pots in spring or summer

Cages of fine mesh which fit closely over pots

Aphid predators, e.g. ladybirds, ladybird larvae or hoverfly larvae

● Place broad bean plants outside. They will (in most parts of Britain) be rapidly infected by the black bean aphid (*Aphis fabae*).

● Adjust the numbers of aphids per plant to the same level by removing them where necessary. Enclose each plant in a mesh cage. Add 1, 2, 4, or 8 predators of the same species to each cage. There should be at least four cages at each predator density.

● Examine the defences and behaviour of the aphids when they are attacked by their predators (Dixon 1976).

● Record the fluctuations in population size of predator and prey over several generations. It may be necessary to transfer the insects to new broad bean plants from time to time.

● Account for the changes in population size of predator and prey.

Six dragonfly or damselfly nymphs
Six short twigs

● Place equal volumes of pond water in each jar and label jars A–F.

● Put one damselfly nymph and one twig in each jar.

● After ten minutes transfer a specific number of *Daphnia* from the culture to each jar:
Jar A, 5 *Daphnia*
Jar B, 10 *Daphnia*
Jar C, 15 *Daphnia*
Jar D, 20 *Daphnia*
Jar E, 30 *Daphnia*
Jar F, 50 *Daphnia*.

● After 40 minutes count and record the numbers of *Daphnia* which remain in each jar. Pool the class results.

● Plot the average number of *Daphnia* eaten against the original number of *Daphnia* in a jar.

What is the relationship between percentage predation and prey density? Relate the results to the regulation of prey populations by predators in nature.

INVESTIGATION 12

Relationship between predator and prey (See also Roberts and King (1987) *Investigation 32.7*)

Requirements
Six jars or beakers of same size
Wide-mouthed teat pipette
Rich culture of water fleas
　(*Daphnia*)

INVESTIGATION 13

The disc experiment — response of a predator to changes in prey density (See also Wratten and Fry (1980) and Monger (1986) *Investigation 28E*.)

Requirements
Paper 60 × 60 cm ruled in 2 cm squares

225 discs per experiment,
preferably textured on one
surface (plastic tokens, bottle
tops, sandpaper discs, metal
rings or rubber washers)
Stop clock
Graph paper

- Work in pairs.
- The 'recorder' anchors the paper
  on a table top and arranges the
  discs on the paper for each of the
  six treatments (carried out in
  random order) — 9, 16, 25, 50,
  100 and 225 discs per 60 × 60 cm
  area. The 'predator' is
  blindfolded away from the sight
  of the discs and the paper. The
  predator searches for discs for
  two minutes by tapping
  randomly on the sheet of paper
  with one finger. After each
  capture the predator counts to
  three before resuming the search
  and the disc is returned to the
  collecting box.
- Repeat for each prey density and
  plot the numbers picked up
  against the density. Calculate the
  percentage of prey 'captured' at
  each density.

  If this model was correct,
  what would happen to a prey
  population which was increasing
  in numbers? Describe ways in
  which you could alter the model
  to give better density-dependent
  regulation.

INVESTIGATION 14

Ecology of annual meadow grass (*Poa annua*)

Requirements

Bare weedy soil, grassland
Quadrat, gridded with string into
5 × 5 cm squares, with thin legs
of steel or wooden dowelling
Hollow steel pegs
Measuring tape

- Examine mapped plants at
  fortnightly intervals. Compare
  dug garden sites with grassland
  sites. Proceed as follows:
- Lay out the quadrat. On the first
  occasion, mark the corner with
  pegs hammered into the ground
  flush with the surface. Measure
  the distances of at least two of
  the pegs from fixed objects such
  as trees so that the quadrat can
  be located again.
- Map on graph paper the
  positions of annual meadow
  grass (AMG) plants in the
  quadrat, noting the numbers of
  tillers and inflorescences on each
  plant.
- Repeat the recording at regular
  intervals of time. Calculate death
  rate, age of death, and rate of
  tiller and inflorescence
  production.

  Races of AMG in grasslands
  tend to be longer lived and
  creeping. They root at the nodes
  and delay inflorescence
  production (Law, Bradshaw and
  Putwain 1977, Law 1979).

- Estimate the number of viable seeds per unit area of soil. Collect soil samples of known area to a depth of 10–15 cm.
- Spread soil out on shallow trays in a light warm place. Keep samples moist.
- Count and remove AMG seedlings and churn up the soil every 40 days.

  Roberts and Feast (1973) describe experiments on the longevity of AMG seeds in soil. Chancellor's (1966) book illustrates most weed seedlings.

- Determine seed output by multiplying the mean number of seeds per inflorescence by the mean number of inflorescences per plant.
- Compare results from disturbed sites and grassland sites.

References to Poa annua:
  R. Law, A. D. Bradshaw, P. D. Putwain. Life history variation in Poa annua, *Evolution* **31**, 233–246, 1977.
  R. Law. The cost of reproduction in annual meadow grass, *American Naturalist* **113**, 3–16, 1979.
  H. A. Roberts, P. M. Feast. Emergence and longevity of seeds of annual weeds in cultivated and undisturbed soil, *Journal of Applied Ecology* **10**, 133–43, 1973.

INVESTIGATION 15

The energetics of a stick insect

(This valuable experiment is discussed in detail by R. J. Slatter, The energy budget of the stick insect. *School Science Review* **62** (219), 312–316, December 1980. It also appears, with reference to the Indian stick insect (*Carausius morosus*), in Monger (1986) *Investigation 29B*, and Roberts and King (1987) *Investigation 32.8*. A similar technique for the Australian stick insect, but without the respirometer, is described in ABAL Unit 9, *Ecology*, 9–11 (Cambridge University Press, 1985).)

Requirements
Bell jar or coffee jar
Petri dish base
McCartney bottles or small beakers
Parafilm or clingfilm
Paper towel or cotton wool
Plasticine
Graph paper
Balance (preferably weighing to
  three decimal places)
Drying oven
Cavity microscope slides
Respirometer
Privet (*Ligustrum ovalifolium*)
  shoots
Indian stick insect (*Carausius
  morosus*)

The object is to measure all four terms in the energy budget equation, in terms of kilojoules, for a stick insect kept for a week on a privet shoot. The equation if $FE = P + R + E$, where FE is the energy in the food eaten, P is energy in production (dry mass of insect, cast skins, eggs), R is energy lost in metabolism (respiration) and E is energy lost in the dry mass of egesta.

- Before setting up the apparatus (figure 97) weigh the insect and the privet shoot. Also trace the outlines of all the leaves onto squared paper, without detaching them from the shoot.
- After a week, record mass of food eaten by (a) reweighing privet shoot and (b) drawing round outlines of leaves again; calculate area by counting squares, convert to mass after weighing leaves of known area. Assume that 1 g fresh privet contains 6 kJ energy.
- After a week, reweigh the insect together with cast skins and eggs. Assume that 1 g of insect production (wet mass) contains 10 kJ.
- After a week, collect all egesta with forceps on to a weighed cavity slide. Reweigh after drying for 48 hours. Assume 1 g dry egesta contains 13 kJ.
- Put insect inside a respirometer and determine its oxygen uptake per minute. Calculate its oxygen uptake per week. Calculate the energy lost in metabolism in a week by assuming that 1 cm of oxygen causes the release of 0.02 kJ heat.
- Assemble the energy budget equation.

    Does it balance, more or

FIGURE 97

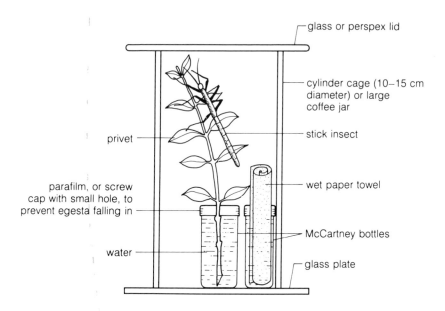

Apparatus suitable for the investigation of the energetics of a stick insect. (Monger, 1986.)

less? Make a list of the possible sources of error. How does the budget differ in insects of different sizes? Why? How might the energy budget of the carnivorous insect differ?

INVESTIGATION 16

Heavy metal or salt tolerance in grasses

(See also Roberts and King (1987) *Investigation 35.3*)

The object is to compare 'average' varieties of a species with varieties taken from spoil heaps (heavy metals) or roadside verges (salt or lead tolerance?) to determine whether nutrient stress has caused the evolution of tolerance (see Bradshaw and McNeilly 1981)

Requirements

Plastic sandwich boxes or beakers
Hoagland's nutrient solution
Alkathene beads
Aquarium pump, rubber tubing and pipette
Stock solutions of sodium chloride or salts of copper, zinc, lead or nickel, to be diluted if necessary
Cling film
'Average', and possibly tolerant clumps of grasses of the same species (obtained from roadsides, salt marshes, spoil heaps or serpentine soils). Suitable species include bent grass (*Agrostis stolonifera*), sweet vernal grass (*Anthoxanthum odoratum*), red fescue (*Festuca rubra*) and cocksfoot (*Dactylis glomerata*). Seeds of the tolerant *F. rubra*

varieties Dawson and Merlin are obtainable from Philip Harris Biological Ltd.

The experiment can be done with either shoots or seeds.

- Add Hoagland's solution to the plastic beakers and add two layers of alkathene beads to each beaker.
- Plant ten shoots of the 'average' variety in half the beakers and ten shoots of the 'tolerant' variety in the other half.
- After 7–10 days, measure and record the lengths in mm of the longest root on each seedling. Return the seedlings to their beakers.
- Repeat after a further four days. Calculate the average increase in root length (mm) in each beaker.
- Add the metal ion or salt solutions. After four more days, measure and record again the lengths of the longest roots. Calculate the mean increase in root length (mm) in each beaker.
- Express the tolerance of the seedlings in each beaker as $a/b \times 100$, where a = mean increase in root length in the stressed solution and b = mean increase in root length in the unstressed solution.

Is there any evidence of tolerance? Of what commercial value might this tolerance be? Suggest how it might have evolved.

INVESTIGATION 17

Biochemical oxygen demand and the fauna of polluted rivers

Requirements

River or stream with a discharge into it of water from a factory or sewage works
Plankton nets
White plastic bowls
Water samples
Dilution water
Chemicals for Winkler method of oxygen determination

● Sample algae and animals at mapped points both upstream and downstream of effluent outflows.

    The animals are sensitive indicators of levels of pollution (figure 98). For identification see Clegg (1980), Mellanby (1963) or Quigley (1977). The B.O.D. is the volume of oxygen in cm^3 absorbed by the organisms in a litre of water incubated for five days at 20°C. Measure the B.O.D. by the method described by McLusky (1971).

● Collect a water sample from each sample site with the least possible disturbance. Add to each sample a known volume of aerated 'dilution water'. This contains nutrient salts so that bacterial activity will not be inhibited during the incubation by lack of nutrient ions.

● Pour each solution into a 250 ml sample bottle. Incubate at 20°C for five days.

FIGURE 98

Class of waterway	Fauna	Biochemical oxygen demand ($mg\ O_2$ absorbed 1^{-1} water at 20°C in 5 days)	Waterway used for:
I	Diverse; salmon, trout, grayling, stonefly and mayfly nymphs, caddis larvae, *Gammarus*.	0–3	Domestic supply
II	Trout rarely dominant; chub, dace, caddis larvae, *Gammarus*.	3–10 (increased in summer at times of low flow)	Agriculture. Industrial processes
III	Roach, gudgeon, *Asellus*; mayfly nymphs and caddis larvae rare.	10–15	Irrigation
IV	Fish absent. Red chironomid larvae (bloodworms) and *Tubifex* worms present.	15–30 (completely deoxygenated from time to time).	Very little. Unsuitable for amenity use
V	Barren, or with fungus or small *Tubifex* worms.	>30	None

Freshwater animals as indicators of pollution. (Adapted from Mills, 1972.)

- Set up control bottles which only contain dilution water.
- Determine the concentration of dissolved oxygen at the beginning and end of the five day period by the Winkler procedure. This involves the titration of a sample of the solution against dilute sodium thiosulphate (McLusky 1971). Alternatively use a calibrated oxygen electrode.
- Relate the B.O.D. of your water samples both to the source of pollution and to the organisms which the river supports at that point.

INVESTIGATION 18

Lichens as indicators of sulphur dioxide pollution

Requirements

Pictures of lichen species (Mabey 1974; Kershaw 1980)

ACE Pollution scale (figure 99)

FIGURE 99

Zone	(Algae), Lichens and mosses	Mean winter SO_2 (μ/m^3)
0	*Pleurococcus* growing on sycamore	Over 170
1	*Lecanora conizaeoides* on trees and acid stone	150–160
2	*Xanthoria parietina* appears on concrete, asbestos and limestone	About 125
3	*Parmelia* appears on acid stone and *Grimmia pulvinata* occurs on limestone or near mortar	About 100
4	Grey leafy (foliose) species (e.g. *Hypogymnia physodes*) start to appear on trees	About 70
5	Shrubby (fruticose) lichens (e.g. *Evernia prunastri*) start to appear on trees	About 40–60
6	*Usnea* becomes abundant	About 35

Pleurococcus is an emerald green alga frequently forming a green slime over moist bark of trees.

Lecanora forms a grey–green crust closely pressed to the bark of living or dead wood, or the surface of stone. Individuals are only 2 mm across.

Xanthoria is red, orange or bright yellow and forms distinct circular or oval patches, lobed at the edge and usually fertile, on walls, roof-tops, trees and fences.

Parmelia species form grey rosettes which are strongly lobed around the margin. They are dark beneath and frequently attached to the substratum, by means of root-like threads. It occurs on stones and trees.

Grimmia pulvinata is a moss which occurs in small, rounded, neat, dense cushions 1–2 cm high, grey with the hair points of the leaves, on rocks and stones.

Hypogymnia physodes is found on trees and fences as grey–green rosettes with the edges strongly lobed. The lobes are 2–4 mm wide and tend to stick up into the air; there are no root-like threads beneath, as in *Parmelia*.

Shrubby lichens are either erect and bush-like, or hanging and tassel-like.

Usnea species are usually grey–green, intricately branched, long, and trailing like long tassels from the branches of trees. They are called 'beard lichens'.

A biological scale, based on the occurrence of *Pleurococcus* and various lichens and mosses, which can be used to estimate the mean winter atmospheric level of sulphur dioxide. The scale is similar to that used in the A.C.E. pollution survey. (Mabey, 1974.)

This exercise can be performed by a class at varying distances and aspects from a possible pollution source (factory, power station, town) or by surveying lichens near home, entering the results on a map of the area and building up a map of local pollution year by year.

- Select a varied area containing trees, walls and houses, with some stone (e.g. gravestones) if possible. Search for the lichen and moss species on the ACE pollution scale, and record those present. The lichen 'zone' depends on the presence of the most pollution-sensitive species recorded.
- Within the sampling area map two or three species with differing pollution tolerances, e.g. *Lecanora conizaeoides* and *Xanthoria parietina*. Note the place, the type of stone or the species of tree on which each lichen occurs. Relate the distribution pattern to traffic flow, factories and the microhabitats suitable for each species.

INVESTIGATION 19

The effect of eutrophication on the growth of duckweed
(See also Roberts and King (1987) *Investigation 32.9*)

Requirements
Beakers (250 cm³ or larger) × 20
Glasshouse
Colorimeter and tubes
Measuring cylinder
Marker for writing on glass
Balance
Sodium nitrate
Calcium monohydrogen phosphate
Duckweed plants (*Lemna* spp.)
Graph paper

- Place 250 cm³ water from the same pond into each of the twenty beakers. Label four beakers C. These are 'control' beakers with no nutrient additions. There are eight nutrient addition treatments, each applied to two beakers. Label two beakers P1, two P2, two N1, two N2, two N1P1, two N1P2, two N2P1, and two N2P2.
In this code: N1 is addition of
1 g sodium nitrate
N2 is addition of 5 g sodium nitrate
P1 is addition of 0.1 g CaHPO₄
P2 is addition of 0.5 g CaHPO₄
- Add the appropriate mass of salts to each beaker and stir thoroughly. To each beaker add ten *Lemna* plants. Count and record the numbers of leaves in each beaker. Arrange the beakers in a glasshouse in a 4 × 5 pattern.
- Examine and count the numbers of leaves every two weeks. Assess by eye the relative

densities of protoctists (algae) in each beaker. After four weeks assess the chlorophyll density in the water in each beaker with a colorimeter.

- Draw graphs of the numbers of leaves per treatment, and the chlorophyll absorption in each treatment, with time.

Was the nitrate or the phosphate the more effective at promoting growth? Together, did they increase growth more than might have been expected from the effects of nitrates or phosphates alone? What are the implications of your results for freshwater habitats?

References

GENERAL REFERENCES
P. A. Colinvaux.
Introduction to Ecology, 3rd Edn.
(Wiley, 1986).

P. R. Ehrlich and A. H. Ehrlich.
Population, Resources, Environment.
(Freeman, 1972).

C. S. Elton.
Animal Ecology. (Sidgewick and
Jackson, 1927).

C. J. Krebs.
*Ecology: the Experimental Analysis of
Distribution and Abundance*, 3rd Edn.
(Harper and Row, 1985).

E. P. Odum.
Fundamentals of Ecology, 3rd Edn.
(Saunders, 1971).

D. F. Owen.
What is Ecology? (Oxford University
Press, 1974).

R. E. Ricklefs.
Ecology. (Nelson, 1973).

M. K. Sands.
Problems in Ecology. (Mills and Boon,
1978).

REFERENCES CHAPTER I
C. S. Elton.
The Pattern of Animal Communities.
(Methuen, 1966).

REFERENCES CHAPTER 2
J. L. Harper.
The Population Biology of Plants.
(Academic Press, 1977).

G. E. Hutchinson.
Concluding remarks. *Cold Spring
Harbour Symposium on Quantitative
Biology* 22, 415–427, 1958.

REFERENCES CHAPTER 3
R. M. Crawford.
The control of anaerobic respiration as
a determining factor in the distribution
of the genus Senecio. *Journal of Ecology*
54, 403–413, 1966.

C. S. Elton.
*The Ecology of Invasions by Animals and
Plants.* (Methuen, 1958).

J. R. Etherington.
*Plant Physiological Ecology. Studies in
Biology 98.* (Arnold, 1978).

J. K. Holloway.
Weed control by insect. *Scientific
American* 197, 56–62, July 1957.

D. H. Janzen.
Coevolution of mutualism between ants
and Acacias in Central America.
Evolution 20, 249–275, 1966.

M. H. Martin.
Conditions affecting the distribution of
Mercurialis perennis L. in certain
Cambridgeshire woods. *Journal of
Ecology* 56, 777–793, 1968.

G. O'Hare.
Soils, Vegetation, Ecosystems (Oliver and
Boyd, 1988).

C. D. Pigott.
The response of plants to climate and climatic change. *In* F. Perring (ed.) *The Flora of a Changing Britain, B. S. B. I. Conference Report* **11**, 32–44, 1970.

J. C. Taylor.
The introduction of exotic plant and animal species into Britain. *Biologist* **26**(5), 229, 1979.

REFERENCES CHAPTER 4
J. A. Gulland. The effect of exploitation on the numbers of marine animals. *Proceedings for the Advance Study Institute for Dynamics Numbers Population* (Oosterbeek, 1970), 450–468.

J. Hudson and A. Watson.
The red grouse. *Biologist* **32**(1), 13–18, 1985.

D. H. Janzen.
Ecology of Plants in the Tropics. Studies in Biology 58. (Arnold, 1975).

J. R. Krebs.
Regulation of numbers in the great tit (*Aves: passeriformes*). *Journal of Zoology* **162**, 317–333, 1970.

D. Lack.
The Natural Regulation of Animal Numbers. (Oxford University Press, 1954).

C. M. Perrins.
British Tits. (Collins, 1979).

C. M. Perrins.
The great tit Parus major. *Biologist* **27** (2), 73–80, 1980.

H. N. Southern.
The natural control of a population of tawny owls (*Strix aluco*). *Journal of Zoology* **162**, 197, 1970.

H. F. van Emden.
Pest Control and its Ecology. Studies in Biology 50. (Arnold, 1974).

REFERENCES CHAPTER 5
J. Gribbin.
Making a date with radiocarbon. *New Scientist* **82**, 532–534, 1979.

D. S. Ranwell.
Ecology of Salt Marshes and Sand Dunes. (Chapman and Hall, 1972).

REFERENCES CHAPTER 6
J. M. Anderson.
Ecology for Environmental Sciences: Biosphere, Ecosystems and Man. (Arnold, 1981).

J. Gribbin.
Woodman, spare that tree. *New Scientist* **81**, 1016–18, 1979.

J. Gribbin.
The greenhouse effect. 'Inside Science'. *New Scientist*, 22 October 1988.

G. E. Likens, F. H. Bormann, N. M. Johnson, D. W. Fisher and R. S. Pierce.
The effect of forest cutting and herbicide treatment on nutrient budgets in the Hubbard Brook watershed–ecosystem. *Ecological Monographs* **40**, 23–47, 1970.

K. G. Porter.
The plant–animal interface in freshwater ecosystems. *American Scientist* **65**, 159–170, 1977.

R. Revelle.
Carbon dioxide and world climate. *Scientific American* **247**, 247(2), 1982.

R. H. Whittaker.
Communities and Ecosystems, 2nd Edn. (Macmillan, 1975).

Sources for figure 71

(*a*) *See figure 67.*

(*b*) from data in H. T. Odum (1970). Summary: an emerging view of the ecological system at El Verde. pp. 191–289 in H. T. Odum, R. F. Pigeon, (eds) *A Tropical Rain Forest. A study of irradiation and ecology at El Verde, Puerto Rico.* (Division of Technical Information, American Atomic Energy Commission.)

(c) from data in L. C. Bliss (1975).
Devon Island, Canada, in T. Rosswall,
O. W. Heal (eds) Structure and
function of tundra ecosystems. *Ecology
Bulletin* (Stockholm) **20**, 17–60.

Sources for figure 72

(a) T. Persson and U. Lohm (1975).
Energetical significance of the annelids
and arthropods in a Swedish grassland
soil. *Ecology Bulletin*. (Stockholm) **23**,
1–211.

(b) see figure 7.

(c) D. J. Crisp (1975).
Secondary productivity of the sea. pp.
71–89 in D. E. Reichie, J. F. Franklin
and D. W. Goodall (eds) *Productivity of
world ecosystems*. (N.A.S. Seattle,
Washington.)

(d) adapted from S. B. Chapman, J.
Hibbie & C. R. Rafarei (1975).
Net aerial production by *Calluna
vulgaris* on lowland heath in Britain.
Journal of Ecology **63**, 233–58.

(e) H. T. Odum (1970) see figure 71.

(f) adapted from R. H. Whittaker
(1970).
Communities and Ecosystems.
(Macmillan).

(g) Open University (1974). S323
course, Unit 5, p. 29, work by O. W.
Heal and R. A. H. Smith.

(h) adapted from J. D. Ovington & D.
Heitkamp (1960).
The accumulation of energy in forest
plantations in Britain. *Journal of
Ecology* **48**, 639–46, and A. MacFadyan
(1964). Grazing in terrestrial and
marine environments. *In*: D. J. Crisp
(ed.) *Symposium of the British Ecological
Society* **4**, 3–20.

REFERENCES CHAPTER 7
J. R. Echols.
Population *vs.* the Environment: a
crisis of too many people. *American
Scientist* **64**, 165–173, 1976.

D. R. Gwatkin and S. K. Brandel.
Life expectancy and population growth
in the third world. *Scientific American*
246(5), 33, 1982.

REFERENCES CHAPTER 8
J. N. B. Bell.
Recent developments in acid rain
research. *Journal of Biological
Education* **22**, (2), 93–98, 1988.

D. L. Hawksworth and F. Rose.
*Lichens as Pollution Monitors. Studies in
Biology 66.* (Arnold, 1976).

J. Lee.
Acid Rain. *Biological Sciences Review* **1**
(1), 15–18, 1988.

K. Mellanby.
*The Biology of Pollution. Studies in
Biology 38.* (Arnold, 1972).

J. H. Ottoway.
*The Biochemistry of Pollution. Studies in
Biology 123.* (Arnold, 1983).

REFERENCES CHAPTER 9
E. Duffey.
Nature Reserves and Wildlife.
(Heinemann, 1974).

C. Pye-Smith and C. Rose.
*Crisis and Conservation: Conflict in the
British Countryside.* (Penguin, 1984).

E. C. Wolf.
Avoiding a mass extinction of species.
In: L. Brown (ed.) *State of the World*
(Norton, 1988), pp.101–117.

REFERENCES PRACTICAL
AIDGAP keys (Field Studies Council
and Richmond Publishing Company,
1975–88) on British water plants,
brown seaweeds, diatoms, diptera,
crabs, slugs, coleoptera, bees, ants and
wasps, grasses, red seaweeds, sea
spiders, freshwater invertebrates,
terrestrial invertebrates, water beetles.

N. T. J. Bailey.
Statistical Methods in Biology, 2nd Edn.
(English Universities Press, 1981).

J. Barrett and C. M. Yonge.
Pocket Guide to the Sea Shore. (Collins, 1958)

G. E. Beedham.
Identification of the British Mollusca. (Hulton Educational Publications, 1973).

H. Belcher and E. Swale.
A Beginner's Guide to Freshwater Algae. (HMSO, 1977).

H. Belcher and E. Swale.
An Illustrated Guide to River Phytoplankton. (HMSO, 1980).

D. P. Bennett and D. A. Humphries.
Introduction to Field Biology, 3rd Edn. (Arnold, 1980).

O. N. Bishop.
Statistics for Biology. (Longman, Microcomputer Edition, 1983).

A. D. Bradshaw and T. McNeilly.
Evolution and Pollution. Studies in Biology 130. (Arnold, 1981).

J. Brodie.
Grassland Studies. (Allen and Unwin, 1985).

R. C. Campbell.
Statistics for Biologists, 2nd Edn. (Cambridge University Press, 1974).

R. J. Chancellor.
The Identification of Weed Seedlings of Farm and Garden. (Blackwell Scientific Publications, 1966).

J. T. Clark.
Stick and Leaf Insects. (Barry Shurlock, 1974).

J. Clegg.
The Observer's Book of Pond Life, 3rd Edn. (Warne, 1980).

M. Chinery.
The Natural History of the Garden. (Collins, 1977).

M. Chinery.
A Field Guide to the Insects of Britain and Northern Europe, 2nd Edn. (Collins, 1976).

F. Clegg.
Simple Statistics. (Cambridge Educational, 1982).

J. L. Cloudsley-Thompson and J. Sankey.
Land Invertebrates. (Methuen, 1961).

M. Collins.
Urban Ecology. (Cambridge University Press, 1984).

S. A. Corbet and R. H. L. Disney (eds).
Naturalists' Handbooks (Richmond Publishing Co. Slough, 1981–90). Separate handbooks on grasshoppers, insects and thistles, solitary wasps, insects on nettles, hoverflies, bumblebees etc.

J. Cousens.
An Introduction to Woodland Ecology. (Oliver and Boyd, 1974).

A. Darlington.
The Ecology of Walls. (Heinemann Educational, 1981).

A. F. G. Dixon.
Biology of Aphids. Studies in Biology 44. (Arnold, 1973).

W. H. Dowdeswell.
Ecology: Principles and Practice. (Heinemann Educational, 1984).

S. M. Evans and J. M. Hardy.
Seashore and Sand Dunes. (Heinemann, 1970).

J. H. R. Gee.
Freshwater Studies. (Allen and Unwin, 1986).

D. L. Hawksworth and F. Rose.
Lichens as Pollution Monitors. Studies In Biology 66. (Arnold, 1976).

D. Hodges.
The Pollen Loads of the Honeybee. (Bee Research Organisation, London, 1964).

M. Jenkins.
Seashore Studies. (Allen and Unwin, 1983).

W. Keble Martin.
The Concise Flora in Colour, 2nd Edn.
(Ebury Press and Michael Joseph,
1965).

M. D. Kerney and R. A. D. Cameron.
*Field Guide to the Land Snails of Britain
and North West Europe.* (Collins, 1979).

K. A. Kershaw.
The Observer's Book of British Lichens,
2nd Edn. (Warne, 1980).

A. Leadley-Brown.
Ecology of Soil Organisms. (Heinemann,
1978).

A. Leadley-Brown.
Freshwater Ecology. (Heinemann,
1986).

T. Lewis and L. R. Taylor.
Introduction to Experimental Ecology.
(Academic Press, 1967).

R. J. Lincoln and J. G. Sheals.
*Invertebrate Animals: Collection and
Preservation.* (British Museum and
Cambridge University Press, 1979).

R. Mabey.
The Pollution Handbook. The ACE/
Sunday Times Clean Air and Water
Surveys. (Penguin, 1974).

D. S. McLusky.
Ecology of Estuaries. (Heinemann,
1971).

H. Mellanby.
Animal Life in Fresh Water, 6th Edn.
(Methuen, 1963).

D. H. Mills.
An Introduction to Freshwater Ecology.
(Oliver and Boyd, 1972).

G. Monger (ed.).
*Revised Nuffield A level Biology,
Practical Guide 7, Ecology.* (Longman,
1986).

Nuffield Advanced Biological Science.
Key to Pond Organisms. (Penguin,
1970).

R. E. Parker.
*Introductory Statistics for Biology.
Studies in Biology 43.* (Arnold, 1973).

K. Paviour-Smith and J. B. Whittaker.
*A Key to the Major Groups of British
Free-living Terrestrial Invertebrates.*
(Blackwell Scientific Publications,
1968).

M. C. F. Proctor and P. F. Yeo.
The Pollination of Flowers. (Collins,
1973).

M. Quigley.
*Invertebrates of Streams and Rivers. A
Key to Identification.* (Arnold, 1977).

M. Quigley (ed.).
Blackwell Habitat Guides. (1986–88),
on flowering plants of salt marshes and
sand dunes, herbaceous flowering
aquatic plants, invertebrate animals of
freshwater, animals and plants of rocky
shores, common plants of woodland,
and land invertebrates.

M. B. V. Roberts and T. J. King.
*Biology — a Functional Approach.
Students' Manual*, 2nd Edn. (Nelson,
1987).

D. Slingsby and C. Cook.
Practical Ecology. (Macmillan, 1986).

D. Smith.
Urban Ecology. (Allen and Unwin,
1984).

S. L. Sutton.
Woodlice. (Pergamon, 1980).

M. A. Wedgwood.
Tackling Biology Projects. (Macmillan/
Institute of Biology, 1987).

G. M. Williams.
Techniques and Fieldwork in Ecology.
(Bell and Hyman, 1987).

S. D. Wratten and G. L. A. Fry.
*Field and Laboratory Exercises in
Ecology.* (Arnold, 1980).

Index